I0044669

# Materials Science and Metallurgy

# Materials Science and Metallurgy

Edited by
**Maddox Moran**

Larsen & Keller
www.larsen-keller.com

Materials Science and Metallurgy
Edited by Maddox Moran
ISBN: 978-1-63549-183-8 (Hardback)

© 2017 Larsen & Keller

目 Larsen & Keller

Published by Larsen and Keller Education,
5 Penn Plaza,
19th Floor,
New York, NY 10001, USA

**Cataloging-in-Publication Data**

Materials Science and Metallurgy / edited by Maddox Moran.
    p. cm.
Includes bibliographical references and index.
ISBN 978-1-63549-183-8
1. Metallurgy. 2. Materials science. 3. Metal-work.
I. Moran, Maddox.
TN607 .M38 2017
669--dc23

This book contains information obtained from authentic and highly regarded sources. All chapters are published with permission under the Creative Commons Attribution Share Alike License or equivalent. A wide variety of references are listed. Permissions and sources are indicated; for detailed attributions, please refer to the permissions page. Reasonable efforts have been made to publish reliable data and information, but the authors, editors and publisher cannot assume any responsibility for the vailidity of all materials or the consequences of their use.

Trademark Notice: All trademarks used herein are the property of their respective owners. The use of any trademark in this text does not vest in the author or publisher any trademark ownership rights in such trademarks, nor does the use of such trademarks imply any affiliation with or endorsement of this book by such owners.

The publisher's policy is to use permanent paper from mills that operate a sustainable forestry policy. Furthermore, the publisher ensures that the text paper and cover boards used have met acceptable environmental accreditation standards.

Printed and bound in the United States of America.

For more information regarding Larsen and Keller Education and its products, please visit the publisher's website www.larsen-keller.com

# Table of Contents

# Preface

This book outlines the processes and applications of metallurgy and metal science in detail. It covers all the important topics of this area. Metallurgy refers to the study of the physical and chemical behavior of metals and their interactions with each others and also the forming and function of alloys. This text presents the complex subject of metal science in the most comprehensible and easy to understand language. The various sub-fields along with the technological progress that have future implications are glanced at in this book. It is a valuable compilation of topics, ranging from the basic to the most complex theories and principles in the field of metallurgy. This textbook is meant for students who are looking for an elaborate reference text on metallurgy and metal science.

A short introduction to every chapter is written below to provide an overview of the content of the book:

**Chapter 1** - Metallurgy is a domain of materials science and engineering, and studies the physical and chemical behavior of metals, metallic compounds and alloys. It also studies the processes and technology that are used in the extraction of metals. This chapter provides a comprehensive overview of metallurgy and metallography- the microscopic study of the physical structure and components of metals; **Chapter 2** - This section illustrates the various branches of metallurgy including pyrometallurgy and extractive metallurgy. These branches seek to purify the ore that has been extracted and are utilized in the production of pure metals. The reader is equipped with information about the various processes involved in each of these branches; **Chapter 3** - Metals are extracted and shaped by various processes and this section details processes like mining, casting, melting, welding, annealing, calcination, recovery, roasting, molding, electrowinning, electroplating etc. The chapter provides information that gives the reader a thorough grasp of the various processes of metallurgy; **Chapter 4** - This chapter studies the various metal alloys in great detail and also has a section dedicated to non-ferrous metals. Alloys are manufactured as they provide better corrosion resistance as well as mechanical as well as torsion strength. Readers are provided with information regarding the formation and uses of the alloys; **Chapter 5** - Due to its tremendous potential of use, iron is utilized in a wide variety of applications. This chapter details the various processes involved in the extraction of iron, namely smelting, etc. The content also has a section about the various furnaces used in the extraction of iron and ironworks. The chapter on ironworks offers an insightful focus, keeping in mind the complex subject matter; **Chapter 6** - Steel is an alloy of iron that has high tensile strength and can be produced at low costs. This chapter describes the modern steel making process. A section of the chapter illustrates the workings of a steel mill and its functions. The chapter strategically encompasses and incorporates the major components and key concepts of steelmaking, providing a complete understanding; **Chapter 7** - Corrosion occurs when metals like iron are exposed to the atmosphere. The metals form stable compounds by chemical and/or electrochemical reactions. The chapter describes in detail the process of anaerobic corrosion which takes place in the presence of anoxic water. The content also talks about cyclic corrosion testing, a test under which samples are subject in laboratories to varying climates that they might encounter in the real world; **Chapter 8** - The process by which various parts, machinery, jewelry, engines and assemblies are created from metals is known as metalworking. The chapter details the processes of casting, rolling, hot working, sheet metal etc. The reader is introduced to the workings of a foundry and the processes involved in cold formed steel. This chapter is a compilation of the various branches of metalworking that form an integral part of the broader subject matter.

I extend my sincere thanks to the publisher for considering me worthy of this task. Finally, I thank my family for being a source of support and help.

**Editor**

# Introduction to Metallurgy

Metallurgy is a domain of materials science and engineering, and studies the physical and chemical behavior of metals, metallic compounds and alloys. It also studies the processes and technology that are used in the extraction of metals. This chapter provides a comprehensive overview of metallurgy and metallography- the microscopic study of the physical structure and components of metals.

## Metallurgy

Georgius Agricola, author of *De re metallica*, an important early work on metal extraction

Metallurgy is a domain of materials science and engineering that studies the physical and chemical behavior of metallic elements, their intermetallic compounds, and their mixtures, which are called alloys. Metallurgy is also the technology of metals: the way in which science is applied to the production of metals, and the engineering of metal components for usage in products for consumers and manufacturers. The production of metals involves the processing of ores to extract the metal they contain, and the mixture of metals, sometimes with other elements, to produce alloys. Metallurgy is distinguished from the craft of metalworking, although metalworking relies on metallurgy, as medicine relies on medical science, for technical advancement.

Smelting Gold in Nicaragua in the La Luz Gold Mine in Siuna and Bonanza about 1959.
Smelting is a basic step in obtaining useable quantities of most metals.

Metallurgy is subdivided into *ferrous metallurgy* (sometimes also known as black metallurgy) and *non-ferrous metallurgy* or *colored metallurgy*. Ferrous metallurgy involves processes and alloys based on iron while non-ferrous metallurgy involves processes and alloys based on other metals. The production of ferrous metals accounts for 95 percent of world metal production.

Pouring Smelted Gold into an ingot at the La Luz Gold Mine in Siuna, Nicaragua about 1959.

## Etymology

The word was originally an alchemist's term for the extraction of metals from minerals, the ending *-urgy* signifying a process, especially manufacturing: it was discussed in this sense in the 1797 Encyclopaedia Britannica. In the late 19th century it was extended to the more general scientific study of metals, alloys, and related processes.

In English, the pronunciation is the more common one in the UK and Common-wealth. The pronunciation is the more common one in the USA, and is the first-listed variant in various American dictionaries (e.g., *Merriam-Webster Collegiate*, *American Heritage*).

## History

Gold headband from Thebes 750–700 BC

The earliest recorded metal employed by humans appears to be gold which can be found free or "native." Small amounts of natural gold have been found in Spanish caves used during the late Paleolithic period, *c.* 40,000 BC. Silver, copper, tin and meteoric iron can also be found in native form, allowing a limited amount of metalworking in early cultures. Egyptian weapons made from meteoric iron in about 3000 BC were highly prized as "daggers from heaven."

Certain metals, notably tin, lead and (at a higher temperature) copper, can be recovered from their ores by simply heating the rocks in a fire or blast furnace, a process known as smelting. The first evidence of this extractive metallurgy dates from the 5th and 6th millennium BC and was found in the archaeological sites of Majdanpek, Yarmovac and Plocnik, all three in Serbia. To date, the earliest evidence of copper smelting is found at the Belovode site, including a copper axe from 5500 BC belonging to the Vinča culture. Other signs of early metals are found from the third millennium BC in places like Palmela (Portugal), Los Millares (Spain), and Stonehenge (United Kingdom). However, the ultimate beginnings cannot be clearly ascertained and new discoveries are both continuous and ongoing.

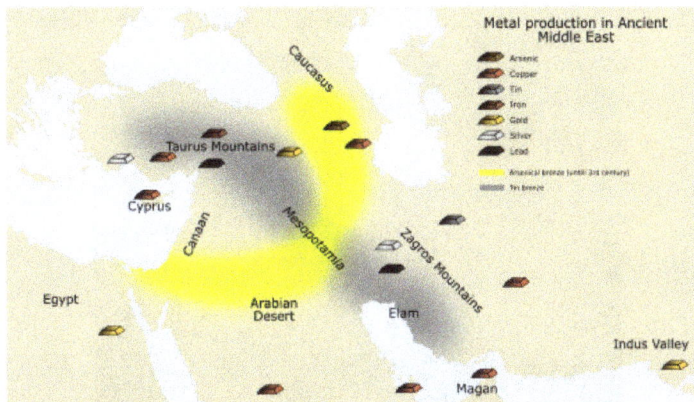

Mining areas of the ancient Middle East. Boxes colors: arsenic is in brown, copper in red, tin in grey, iron in reddish brown, gold in yellow, silver in white and lead in black. Yellow area stands for arsenic bronze, while grey area stands for tin bronze.

These first metals were single ones or as found. About 3500 BC, it was discovered that by combining copper and tin, a superior metal could be made, an alloy called bronze, representing a major technological shift known as the Bronze Age.

The extraction of iron from its ore into a workable metal is much more difficult than for copper or tin. The process appears to have been invented by the Hittites in about 1200 BC, beginning the Iron Age. The secret of extracting and working iron was a key factor in the success of the Philistines.

Historical developments in ferrous metallurgy can be found in a wide variety of past cultures and civilizations. This includes the ancient and medieval kingdoms and empires of the Middle East and Near East, ancient Iran, ancient Egypt, ancient Nubia, and Anatolia (Turkey), Ancient Nok, Carthage, the Greeks and Romans of ancient Europe, medieval Europe, ancient and medieval China, ancient and medieval India, ancient and medieval Japan, amongst others. Many applications, practices, and devices associated or involved in metallurgy were established in ancient China, such as the innovation of the blast furnace, cast iron, hydraulic-powered trip hammers, and double acting piston bellows.

A 16th century book by Georg Agricola called *De re metallica* describes the highly developed and complex processes of mining metal ores, metal extraction and metallurgy of the time. Agricola has been described as the "father of metallurgy".

## Extraction

Furnace bellows operated by waterwheels, Yuan Dynasty, China.

Aluminium plant in Žiar nad Hronom (Central Slovakia)

Extractive metallurgy is the practice of removing valuable metals from an ore and refining the extracted raw metals into a purer form. In order to convert a metal oxide or sulfide to a purer metal, the ore must be reduced physically, chemically, or electrolytically.

Extractive metallurgists are interested in three primary streams: feed, concentrate (valuable metal oxide/sulfide), and tailings (waste). After mining, large pieces of the ore feed are broken through crushing and/or grinding in order to obtain particles small enough where each particle is either mostly valuable or mostly waste. Concentrating the particles of value in a form supporting separation enables the desired metal to be removed from waste products.

Mining may not be necessary if the ore body and physical environment are conducive to leaching. Leaching dissolves minerals in an ore body and results in an enriched solution. The solution is collected and processed to extract valuable metals.

Ore bodies often contain more than one valuable metal. Tailings of a previous process may be used as a feed in another process to extract a secondary product from the original ore. Additionally, a concentrate may contain more than one valuable metal. That concentrate would then be processed to separate the valuable metals into individual constituents.

## Alloys

Casting bronze

Common engineering metals include aluminium, chromium, copper, iron, magnesium, nickel, titanium and zinc. These are most often used as alloys. Much effort has been placed on understanding the iron-carbon alloy system, which includes steels and cast irons. Plain carbon steels (those that contain essentially only carbon as an alloying element) are used in low-cost, high-strength applications where weight and corrosion are not a problem. Cast irons, including ductile iron, are also part of the iron-carbon system.

Stainless steel or galvanized steel are used where resistance to corrosion is important. Aluminium alloys and magnesium alloys are used for applications where strength and lightness are required.

Copper-nickel alloys (such as Monel) are used in highly corrosive environments and for non-magnetic applications. Nickel-based superalloys like Inconel are used in high-temperature applications such as gas turbines, turbochargers, pressure vessels, and heat exchangers. For extremely high temperatures, single crystal alloys are used to minimize creep.

## Production

In production engineering, metallurgy is concerned with the production of metallic components for use in consumer or engineering products. This involves the production of alloys, the shaping, the heat treatment and the surface treatment of the product. The task of the metallurgist is to achieve balance between material properties such as cost, weight, strength, toughness, hardness, corrosion, fatigue resistance, and performance in temperature extremes. To achieve this goal, the operating environment must be carefully considered. In a saltwater environment, ferrous metals and some aluminium alloys corrode quickly. Metals exposed to cold or cryogenic conditions may endure a ductile to brittle transition and lose their toughness, becoming more brittle and prone to cracking. Metals under continual cyclic loading can suffer from metal fatigue. Metals under constant stress at elevated temperatures can creep.

## Metalworking Processes

Metals are shaped by processes such as:

- casting – molten metal is poured into a shaped mold.

- forging – a red-hot billet is hammered into shape.

- rolling – a billet is passed through successively narrower rollers to create a sheet.

- laser cladding – metallic powder is blown through a movable laser beam (e.g. mounted on a NC 5-axis machine). The resulting melted metal reaches a substrate to form a melt pool. By moving the laser head, it is possible to stack the tracks and build up a three-dimensional piece.

- extrusion – a hot and malleable metal is forced under pressure through a die, which shapes it before it cools.

- sintering – a powdered metal is heated in a non-oxidizing environment after being compressed into a die.

- machining – lathes, milling machines, and drills cut the cold metal to shape.

- fabrication – sheets of metal are cut with guillotines or gas cutters and bent and welded into structural shape.

- 3D printing – Sintering or melting powder metal in a very small point on a moving 'print head' moving in 3D space to make any object to shape.

Cold-working processes, in which the product's shape is altered by rolling, fabrication or other processes while the product is cold, can increase the strength of the product by a process called work hardening. Work hardening creates microscopic defects in the metal, which resist further changes of shape.

Various forms of casting exist in industry and academia. These include sand casting, investment casting (also called the "lost wax process"), die casting, and continuous casting.

## Heat Treatment

Metals can be heat-treated to alter the properties of strength, ductility, toughness, hardness and/or resistance to corrosion. Common heat treatment processes include annealing, precipitation strengthening, quenching, and tempering. The annealing process softens the metal by heating it and then allowing it to cool very slowly, which gets rid of stresses in the metal and makes the grain structure large and soft-edged so that when the metal is hit or stressed it dents or perhaps bends, rather than breaking; it is also easier to sand, grind, or cut annealed metal. Quenching is the process of cooling a high-carbon steel very quickly after heating, thus "freezing" the steel's molecules in the very hard martensite form, which makes the metal harder. There is a balance between hardness and toughness in any steel; the harder the steel, the less tough or impact-resistant it is, and the more impact-resistant it is, the less hard it is. Tempering relieves stresses in the metal that were caused by the hardening process; tempering makes the metal less hard while making it better able to sustain impacts without breaking.

Often, mechanical and thermal treatments are combined in what are known as thermo-mechanical treatments for better properties and more efficient processing of materials. These processes are common to high-alloy special steels, superalloys and titanium alloys.

## Plating

Electroplating is a chemical surface-treatment technique. It involves bonding a thin layer of another metal such as gold, silver, chromium or zinc to the surface of the product. It is used to reduce corrosion as well as to improve the product's aesthetic appearance.

## Thermal Spraying

Thermal spraying techniques are another popular finishing option, and often have better high temperature properties than electroplated coatings.

Metallography allows the metallurgist to study the microstructure of metals.

Metallurgists study the microscopic and macroscopic properties using metallography, a technique invented by Henry Clifton Sorby. In metallography, an alloy of interest is ground flat and polished to a mirror finish. The sample can then be etched to reveal the microstructure and macrostructure of the metal. The sample is then examined in an optical or electron microscope, and the image contrast provides details on the composition, mechanical properties, and processing history.

Crystallography, often using diffraction of x-rays or electrons, is another valuable tool available to the modern metallurgist. Crystallography allows identification of unknown materials and reveals the crystal structure of the sample. Quantitative crystallography can be used to calculate the amount of phases present as well as the degree of strain to which a sample has been subjected.

## Conferences

EMC, the European Metallurgical Conference has developed to the most important networking business event dedicated to the non-ferrous metals industry in Europe. From the start of the conference sequence in 2001 at Friedrichshafen it was host of the most relevant metallurgists from all countries of the world. The European Metallurgical Conference is organized by GDMB Society of Metallurgists and Miners.

# Metallography

Metallography is the study of the physical structure and components of metals, typically using microscopy.

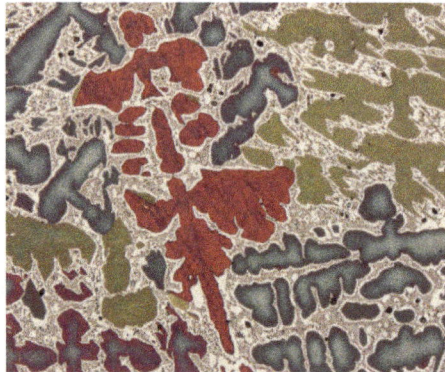

A micrograph of bronze revealing a cast dendritic structure

In some cases, the metallographic structure is large enough to be seen with the unaided eye.

Ceramic and polymeric materials may also be prepared using metallographic techniques, hence the terms ceramography, plastography and, collectively, materialography.

## Preparing Metallographic Specimens

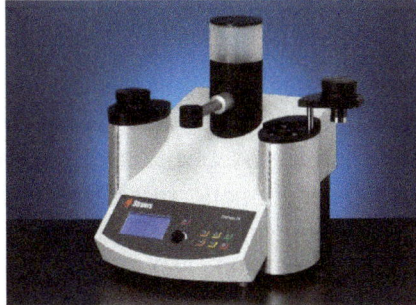

Hot mounting: The specimens are placed in the mounting press, and the resin is added. The specimens are mounted under heat and high pressure.

Cold mounting: The specimens are placed in a mounting cup and mounting material is then poured over the specimens. A vacuum impregnation unit (photo) is used for mounting of porous materials.

The MD-System is an example of a reusable pad for use with diamond suspension. A single magnetic platen is positioned on the grinding and polishing machine to support the preparation pads.

The surface of a metallographic specimen is prepared by various methods of grinding, polishing, and etching. After preparation, it is often analyzed using optical or electron microscopy. Using only metallographic techniques, a skilled technician can identify alloys and predict material properties.

Mechanical preparation is the most common preparation method. Successively finer abrasive particles are used to remove material from the sample surface until the desired surface quality is achieved. Many different machines are available for doing this grinding and polishing, which are able to meet different demands for quality, capacity, and reproducibility.

A systematic preparation method is the easiest way to achieve the true structure. Sample preparation must therefore pursue rules which are suitable for most materials. Different materials with similar properties (hardness and ductility) will respond alike and thus require the same consumables during preparation.

Metallographic specimens are typically "mounted" using a hot compression thermosetting resin. In the past, phenolic thermosetting resins have been used, but modern epoxy is becoming more popular because reduced shrinkage during curing results in a better mount with superior edge retention. A typical mounting cycle will compress the specimen and mounting media to 4,000 psi (28 MPa) and heat to a temperature of 350 °F (177 °C). When specimens are very sensitive to temperature, "cold mounts" may be made with a two-part epoxy resin. Mounting a specimen provides a safe, standardized, and ergonomic way by which to hold a sample during the grinding and polishing operations.

A macro etched copper disc

After mounting, the specimen is wet ground to reveal the surface of the metal. The specimen is successively ground with finer and finer abrasive media. Silicon carbide abrasive paper was the first method of grinding and is still used today. Many metallographers, however, prefer to use a diamond grit suspension which is dosed onto a reusable fabric pad throughout the polishing process. Diamond grit in suspension might start at 9 micrometres and finish at one micrometre. Generally, polishing with diamond suspension gives finer results than using silicon carbide papers (SiC papers), especially with revealing porosity, which silicon carbide paper sometimes "smear" over. After grinding the specimen, polishing is performed. Typically, a specimen is polished with a slurry of alumina, silica, or diamond on a napless cloth to produce a scratch-free mirror finish, free from smear, drag, or pull-outs and with minimal deformation remaining from the preparation process.

After polishing, certain microstructural constituents can be seen with the microscope, e.g., inclusions and nitrides. If the crystal structure is non-cubic (e.g., a metal with a hexagonal-closed packed crystal structure, such as Ti or Zr) the microstructure can be revealed without etching using crossed polarized light (light microscopy). Otherwise, the microstructural constituents of the specimen are revealed by using a suitable chemical or electrolytic etchant.

## Analysis Techniques

Many different microscopy techniques are used in metallographic analysis.

Prepared specimens should be examined with the unaided eye after etching to detect any visible areas that have responded to the etchant differently from the norm as a guide to where microscopical examination should be employed. Light optical microscopy (LOM) examination should always be performed prior to any electron metallographic (EM) technique, as these are more time-consuming to perform and the instruments are much more expensive.

Further, certain features can be best observed with the LOM, e.g., the natural color of a constituent can be seen with the LOM but not with EM systems. Also, image contrast of microstructures at relatively low magnifications, e.g., <500X, is far better with the LOM than with the scanning electron microscope (SEM), while transmission electron microscopes (TEM) generally cannot be utilized at magnifications below about 2000 to 3000X. LOM examination is fast and can cover a large area. Thus, the analysis can determine if the more expensive, more time-consuming examination techniques using the SEM or the TEM are required and where on the specimen the work should be concentrated.

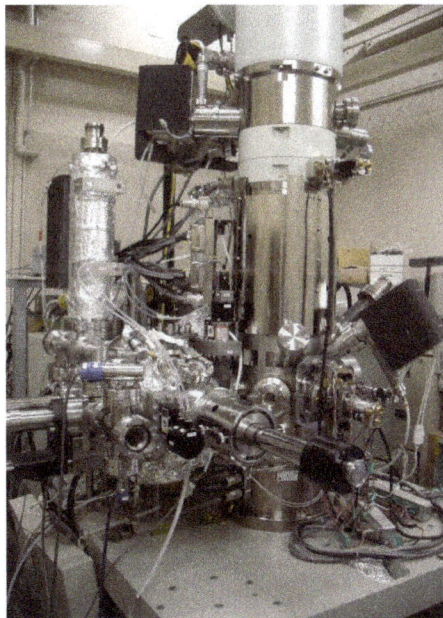

Scanning transmission electron microscope, used in metallography.

## Design, Resolution, and Image Contrast

Light microscopes are designed for placement of the specimen's polished surface on the specimen stage either upright or inverted. Each type has advantages and disadvantages. Most LOM work is done at magnifications between 50 and 1000X. However, with a good microscope, it is possible to perform examination at higher magnifications, e.g., 2000X, and even higher, as long as diffraction fringes are not present to distort the image. However, the resolution limit of the LOM will not be better than about 0.2 to 0.3 micrometers. Special methods are be used at magnifications below 50X, which can be very helpful when examining the microstructure of cast specimens where greater spatial coverage in the field of view may be required to observe features such as dendrites.

Besides considering the resolution of the optics, one must also maximize visibility by maximizing image contrast. A microscope with excellent resolution may not be able to image a structure, that is there is no visibility, if image contrast is poor. Image contrast depends upon the quality of the optics, coatings on the lenses, and reduction of flare and glare; but, it also requires proper specimen preparation and good etching techniques. So, obtaining good images requires maximum resolution and image contrast.

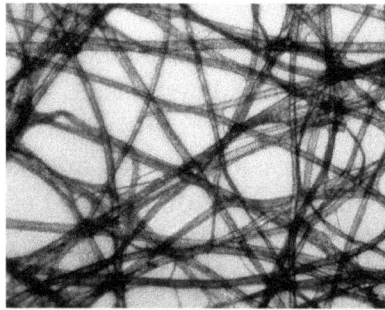

Bright field illumination, where sample contrast comes from absorbance of light in the sample.

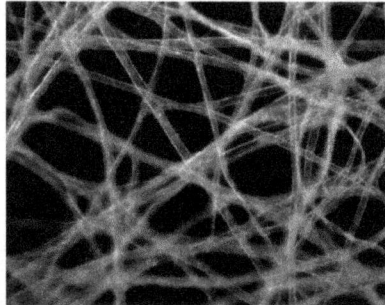

Dark field illumination, sample contrast comes from light scattered by the sample.

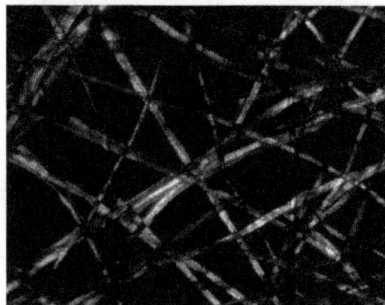

Cross-polarized light illumination, where sample contrast comes from rotation of polarized light through the sample.

## Bright and Dark Field Microscopy

Most LOM observations are conducted using bright field (BF) illumination, where the image of any flat feature perpendicular to the incident light path is bright, or appears to be white. But, other illumination methods can be used and, in some cases, may provide superior images with greater detail. Dark field microscopy (DF), is an alternative method of observation that provides high contrast images and actually greater resolution than bright field. In dark field, the light from features perpendicular to the optical axis is blocked and appears dark while the light from features inclined to the surface, which look dark in BF, appear bright, or "self luminous" in DF. Grain boundaries, for example, are more vivid in DF than BF.

## Polarized Light Microscopy

Polarized light (PL) is very useful when studying the structure of metals with non-cubic crystal structures (mainly metals with hexagonal close-packed (hcp) crystal structures). If the specimen is prepared with minimal damage to the surface, the structure can be seen vividly in cross-polarized light (the optic axis of the polarizer and analyzer are 90 degrees to each other, i.e., crossed). In some cases, an hcp metal can be chemically etched and then examined more effectively with PL. Tint etched surfaces, where a thin film (such as a sulfide, molybdate, chromate or elemental selenium film) is grown epitaxially on the surface to a depth where interference effects are created when examined with BF producing color images, can be improved with PL. If it is difficult to get a good interference film with good coloration, the colors can be improved by examination in PL using a sensitive tint (ST) filter.

## Differential Interference Contrast Microscopy

Another useful imaging mode is differential interference contrast (DIC), which is usually obtained with a system designed by the Polish physicist Georges Nomarski. This system gives the best detail. DIC converts minor height differences on the plane-of-polish, invisible in BF, into visible detail. The detail in some cases can be quite striking and very useful. If an ST filter is used along with a Wollaston prism, color is introduced. The colors are controlled by the adjustment of the Wollaston prism, and have no specific physical meaning, per se. But, visibility may be better.

## Oblique Illumination

DIC has largely replaced the older oblique illumination (OI) technique, which was available on reflected light microscopes prior to about 1975. In OI, the vertical illuminator is offset from perpendicular, producing shading effects that reveal height differences. This procedure reduces resolution and yields uneven illumination across the field of view. Nevertheless, OI was useful when people needed to know if a second phase particle was standing above or was recessed below the plane-of-polish, and is still available on a few microscopes. OI can be created on any microscope by placing a piece of paper under one corner of the mount so that the plane-of-polish is no longer perpendicular to the optical axis.

## Scanning Electron and Transmission Electron Microscopes

If a specimen must be observed at higher magnification, it can be examined with a scanning electron microscope (SEM), or a transmission electron microscope (TEM). When equipped with an energy dispersive spectrometer (EDS), the chemical composition of the microstructural features can be determined. The ability to detect low-atomic number elements, such as carbon, oxygen, and nitrogen, depends upon the nature of the detector used. But, quantification of these elements by EDS is difficult and their minimum detectable limits are higher than when a wavelength-dispersive spectrometer (WDS) is used. But quantification of composition by EDS has improved greatly over time. The WDS system has historically had better sensitivity (ability to detect low amounts of an element) and ability to detect low-atomic weight elements, as well as better quantification of compositions, compared to EDS, but it was slower to use. Again, in recent years, the speed required to perform WDS analysis has improved substantially. Historically, EDS was used with the SEM while WDS was used with the electron microprobe analyzer (EMPA). Today, EDS and WDS is used with both the SEM and the EMPA. However, a dedicated EMPA is not as common as an SEM.

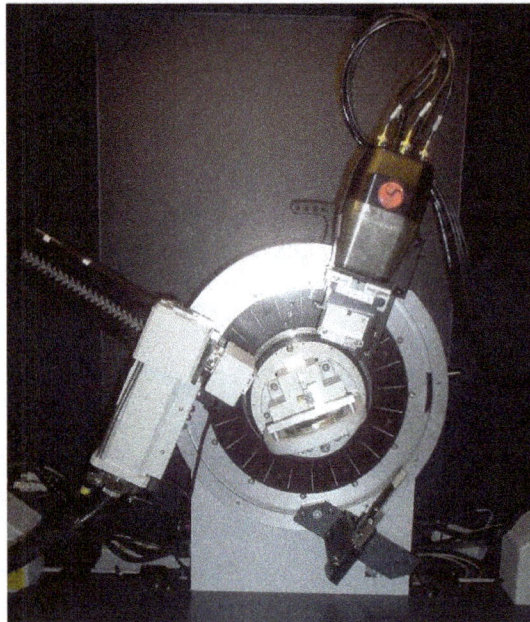

An x-ray diffractometer.

## X-ray Diffraction Techniques

Characterization of microstructures has also been performed using x-ray diffraction (XRD) techniques for many years. XRD can be used to determine the percentages of various phases present in a specimen if they have different crystal structures. For example, the amount of retained austenite in a hardened steel is best measured using XRD (ASTM E 975). If a particular phase can be chemically extracted from a bulk specimen, it can be identified using XRD based on the crystal structure and lattice dimensions. This work can be complemented by EDS and/or WDS analysis where the chemical composition is quantified. But EDS and WDS are difficult to apply to particles less than 2-3 micrometers in diameter. For smaller particles, diffraction techniques can be performed using the TEM for identification and EDS can be performed on small particles if they are extracted from the matrix using replication methods to avoid detection of the matrix along with the precipitate.

## Quantitative Metallography

A number of techniques exist to quantitatively analyze metallographic specimens. These techniques are valuable in the research and production of all metals and alloys and non-metallic or composite materials.

Microstructural quantification is performed on a prepared, two-dimensional plane through the three-dimensional part or component. Measurements may involve simple metrology techniques, e.g., the measurement of the thickness of a surface coating, or the apparent diameter of a discrete second-phase particle, (for example, spheroidal graphite in ductile iron). Measurement may also require application of stereology to assess matrix and second-phase structures. Stereology is the field of taking 0-, 1- or 2-dimensional measurements on the two-dimensional sectioning plane and estimating the amount, size, shape or distribution of the microstructure in three dimensions. These measurements may be made using manual procedures with the aid of templates overlaying the microstructure, or with automated image analyzers. In all cases, adequate sampling must be made to obtain a proper statistical basis for the measurement. Efforts to eliminate bias are required.

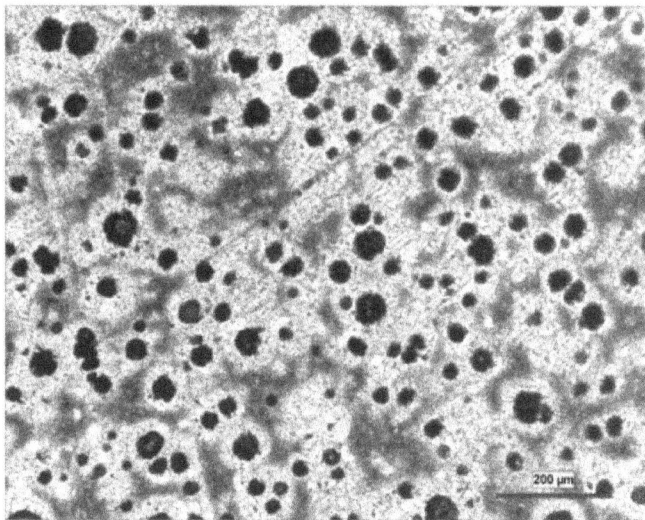

An image of the microstructures of ductile cast iron.

Some of the most basic measurements include determination of the volume fraction of a phase or constituent, measurement of the grain size in polycrystalline metals and alloys, measurement of the size and size distribution of particles, assessment of the shape of particles, and spacing between particles.

Standards organizations, including ASTM International's Committee E-4 on Metallography and some other national and international organizations, have developed standard test methods describing how to characterize microstructures quantitatively.

For example, the amount of a phase or constituent, that is, its volume fraction, is defined in ASTM E 562; manual grain size measurements are described in ASTM E 112 (equiaxed grain structures with a single size distribution) and E 1182 (specimens with a bi-modal grain size distribution); while ASTM E 1382 describes how any grain size type or condition can be measured using image analysis methods. Characterization of nonmetallic inclusions using standard charts is described in ASTM E 45 (historically, E 45 covered only manual chart methods and an image analysis method for making such chart measurements was described in ASTM E 1122. The image analysis methods are currently being incorporated into E 45). A stereological method for characterizing discrete second-phase particles, such as nonmetallic inclusions, carbides, graphite, etc., is presented in ASTM E 1245.

# References

- Arthur Reardon (2011), Metallurgy for the Non-Metallurgist (2nd edition), ASM International, ISBN 978-1-61503-821-3

- Robert K.G. Temple (2007). The Genius of China: 3,000 Years of Science, Discovery, and Invention (3rd edition). London: André Deutsch. pp. 44–56. ISBN 978-0-233-00202-6.

- H.I. Haiko, V.S. Biletskyi. First metals discovery and development the sacral component phenomenon. // Theoretical and Practical Solutions of Mineral Resources Mining // A Balkema Book, London, 2015, p. 227-233..

- E. Photos, E. (2010). "The Question of Meteoritic versus Smelted Nickel-Rich Iron: Archaeological Evidence and Experimental Results" (PDF). World Archaeology 20 (3): 403. doi:10.1080/00438243.1989.9980081. JSTOR 124562.

- Radivojević, Miljana; Rehren, Thilo; Pernicka, Ernst; Šljivar, Dušan; Brauns, Michael; Borić, Dušan (2010). "On the origins of extractive metallurgy: New evidence from Europe". Journal of Archaeological Science 37 (11): 2775. doi:10.1016/j.jas.2010.06.012.

# Branches of Metallurgy

This section illustrates the various branches of metallurgy including pyrometallurgy and extractive metallurgy. These branches seek to purify the ore that has been extracted and are utilized in the production of pure metals. The reader is equipped with information about the various processes involved in each of these branches.

## Extractive Metallurgy

Extractive metallurgy is a branch of metallurgical engineering wherein process and methods of extraction of metals from their natural mineral deposits are studied. The field is a materials science, covering all aspects of the types of ore, washing, concentration, separation, chemical processes and extraction of pure metal and their alloying to suit various applications, sometimes for direct use as a finished product, but more often in a form that requires further working to achieve the given properties to suit the applications.

The field of ferrous and non-ferrous extractive metallurgy have specialties that are generically grouped into the categories of mineral processing, hydrometallurgy, pyrometallurgy, and electrometallurgy based on the process adopted to extract the metal. Several processes are used for extraction of same metal depending on occurrence and chemical requirements.

### Mineral Processing

Mineral processing begins with beneficiation, consisting of initially breaking down the ore to required sizes depending on the concentration process to be followed, by crushing, grinding, sieving etc. Thereafter, the ore is physically separated from any unwanted impurity, depending on the form of occurrence and/or further process involved. Separation processes take advantage of physical properties of the materials. These physical properties can include density, particle size and shape, electrical and magnetic properties, and surface properties. Major physical and chemical methods include magnetic separation, froth floatation, leaching etc., whereby the impurities and unwanted materials are removed from the ore and the base ore of the metal is concentrated, meaning the percentage of metal in the ore is increased. This concentrate is then either processed to remove moisture or else used as is for extraction of the metal or made into shapes and forms that can undergo further processing, with ease of handling.

Ore bodies often contain more than one valuable metal. Tailings of a previous process may be used as a feed in another process to extract a secondary product from the original ore. Additionally, a concentrate may contain more than one valuable metal. That concentrate would then be processed to separate the valuable metals into individual constituents.

## Hydrometallurgy

Hydrometallurgy is concerned with processes involving aqueous solutions to extract metals from ores. The first step in the hydrometallurgical process is leaching, which involves dissolution of the valuable metals into the aqueous solution and /or a suitable solvent. After the solution is separated from the ore solids, the extract is often subjected to various processes of purification and concentration before the valuable metal is recovered either in its metallic state or as a chemical compound. This may include precipitation, distillation, adsorption, and solvent extraction. The final recovery step may involve precipitation, cementation, or an electrometallurgical process. Sometimes, hydrometallurgical processes may be carried out directly on the ore material without any pretreatment steps. More often, the ore must be pretreated by various mineral processing steps, and sometimes by pyrometallurgical processes.

# Pyrometallurgy

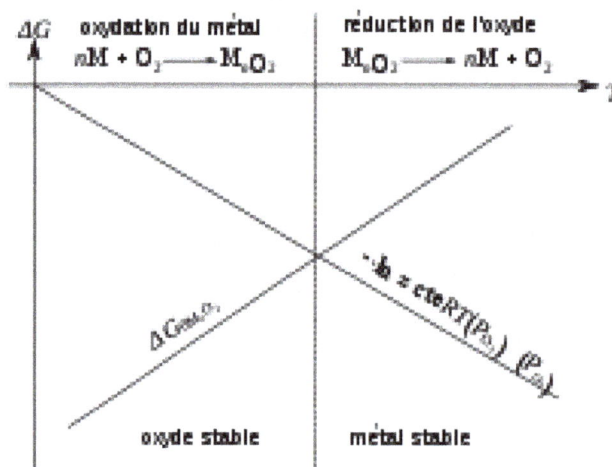

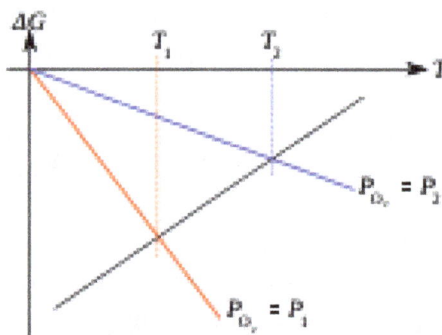

Ellingham diagram for high temperature oxidation

Pyrometallurgy involves high temperature processes where chemical reactions take place among gases, solids, and molten materials. Solids containing valuable metals are treated to form intermediate compounds for further processing or converted into their elemental or metallic state. Pyrometallurgical processes that involve gases and solids are typified by calcining and roasting operations. Processes that produce molten products are collectively referred to as smelting operations. The energy required to sustain the high temperature pyrometallurgical processes may derive from the exothermic nature of the chemical reactions taking place. Typically, these reactions are oxidation, e.g. of sulfide to sulfur dioxide . Often, however, energy must be added to the process by combustion of fuel or, in the case of some smelting processes, by the direct application of electrical energy.

Ellingham Diagrams are a useful way of analysing the possible reactions, and so predicting their outcome.

## Electrometallurgy

Electrometallurgy involves metallurgical processes that take place in some form of electrolytic cell. The most common types of electrometallurgical processes are electrowinning and electro-refining. Electrowinning is an electrolysis process used to recover metals in aqueous solution, usually as the result of an ore having undergone one or more hydrometallurgical processes. The metal of interest is plated onto the cathode, while the anode is an inert electrical conductor. Electro-refining is used to dissolve an impure metallic anode (typically from a smelting process) and produce a high purity cathode. Fused salt electrolysis is another electrometallurgical process whereby the valuable metal has been dissolved into a molten salt which acts as the electrolyte, and the valuable metal collects on the cathode of the cell. The fused salt electrolysis process is conducted at temperatures sufficient to keep both the electrolyte and the metal being produced in the molten state. The scope of electrometallurgy has significant overlap with the areas of hydrometallurgy and (in the case of fused salt electrolysis) pyrometallurgy. Additionally, electrochemical phenomena play a considerable role in many mineral processing and hydrometallurgical processes.

## Pyrometallurgy

Pyrometallurgy is a branch of extractive metallurgy. It consists of the thermal treatment of minerals and metallurgical ores and concentrates to bring about physical and chemical transformations in the materials to enable recovery of valuable metals. Pyrometallurgical treatment may produce products able to be sold such as pure metals, or intermediate compounds or alloys, suitable as feed for further processing. Examples of elements extracted by pyrometallurgical processes include the oxides of less reactive elements like Fe, Cu, Zn, Chromium, Tin, Manganese.

Pyrometallurgical processes are generally grouped into one or more of the following categories:

- Calcining

- Roasting

- Smelting

- Refining

Most pyrometallurgical processes require energy input to sustain the temperature at which the process takes place. The energy is usually provided in the form of combustion or from electrical heat. When sufficient material is present in the feed to sustain the process temperature solely by exothermic reaction (i.e. without the addition of fuel or electrical heat), the process is said to be "autogenous." Processing of some sulfide ores exploit the exothermicity of their combustion.

## Calcination

Calcination is thermal decomposition of a material. Examples include decomposition of hydrates such as ferric hydroxide to ferric oxide and water vapor. The decomposition of calcium carbonate to calcium oxide and carbon dioxide as well as iron carbonate to iron oxide:

$$CaCO_3 \rightarrow CaO + CO_2$$

Calcination processes are carried out in a variety of furnaces, including shaft furnaces, rotary kilns, and fluidized bed reactors.

## Roasting

Roasting consists of thermal gas-solid reactions, which can include oxidation, reduction, chlorination, sulfation, and pyrohydrolysis.

The most common example of roasting is the oxidation of metal sulfide ores. The metal sulfide is heated in the presence of air to a temperature that allows the oxygen in the air to react with the sulfide to form sulfur dioxide gas and solid metal oxide. The solid product from roasting is often called "calcine." In oxidizing roasting, if the temperature and gas conditions are such that the sulfide feed is completely oxidized, the process is known as "dead roasting." Sometimes, as in the case of pre-treating reverberatory or electric smelting furnace feed, the roasting process is performed with less than the required amount of oxygen to fully oxidize the feed. In this case, the process is called "partial roasting," because the sulfur is only partially removed. Finally, if the temperature and gas conditions are controlled such that the sulfides in the feed react to form metal sulfates instead of metal oxides, the process is known as "sulfation roasting." Sometimes, temperature and gas conditions can be maintained such that a mixed sulfide feed (for instance a feed containing both copper sulfide and iron sulfide) reacts such that one metal forms a sulfate and the other forms an oxide, the process is known as "selective roasting" or "selective sulfation."

## Smelting

Smelting involves thermal reactions in which at least one product is a molten phase.

Metal oxides can then be smelted by heating with coke or charcoal (forms of carbon), a reducing agent that liberates the oxygen as carbon dioxide leaving a refined mineral. Concern about the production of carbon dioxide is only a recent worry, following the identification of the enhanced greenhouse effect.

Carbonate ores are also smelted with charcoal, but sometimes need to be calcined first.

Other materials may need to be added as flux, aiding the melting of the oxide ores and assisting in the formation of a slag, as the flux reacts with impurities, such as silicon compounds.

Smelting usually takes place at a temperature above the melting point of the metal, but processes vary considerably according to the ore involved and other matters.

## Refining

Refining is the removal of impurities from materials by a thermal process. This covers a wide range of processes, involving different kinds of furnace or other plant.

The term, 'refining' can also refer to certain electrolytic processes. Accordingly, some kinds of pyrometallurgical refining are referred to as 'fire refining'.

# Processes of Metallurgy

Metals are extracted and shaped by various processes and this section details processes like mining, casting, melting, welding, annealing, calcination, recovery, roasting, molding, electrowinning, electroplating etc. The chapter provides information that gives the reader a thorough grasp of the various processes of metallurgy.

## Mining

Mining is the extraction of valuable minerals or other geological materials from the earth from an orebody, lode, vein, seam, reef or placer deposits which forms the mineralized package of economic interest to the miner.

Surface coal mining

Ores recovered by mining include metals, coal, oil shale, gemstones, limestone, dimension stone, rock salt, potash, gravel, and clay. Mining is required to obtain any material that cannot be grown through agricultural processes, or created artificially in a laboratory or factory. Mining in a wider sense includes extraction of any non-renewable resource such as petroleum, natural gas, or even water.

Simplified world active active mining map

Sulfur miner with 90 kg of sulfur carried from the floor of the Ijen Volcano (2015)

Mining of stones and metal has been a human activity since pre-historic times. Modern mining processes involve prospecting for ore bodies, analysis of the profit potential of a proposed mine, extraction of the desired materials, and final reclamation of the land after the mine is closed.

Mining operations usually create a negative environmental impact, both during the mining activity and after the mine has closed. Hence, most of the world's nations have passed regulations to decrease the impact. Worker safety has long been a concern as well, and modern practices have significantly improved safety in mines.

Levels of metals recycling are generally low. Unless future end-of-life recycling rates are stepped up, some rare metals may become unavailable for use in a variety of consumer products. Due to the low recycling rates, some landfills now contain higher concentrations of metal than mines themselves.

## History

## Prehistoric Mining

Since the beginning of civilization, people have used stone, ceramics and, later, metals found close to the Earth's surface. These were used to make early tools and weapons; for example, high quality flint found in northern France and southern England was used to create flint tools. Flint mines have been found in chalk areas where seams of the stone were followed underground by shafts and galleries. The mines at Grimes Graves are especially famous, and like most other flint mines, are

Neolithic in origin (ca 4000 BC-ca 3000 BC). Other hard rocks mined or collected for axes included the greenstone of the Langdale axe industry based in the English Lake District.

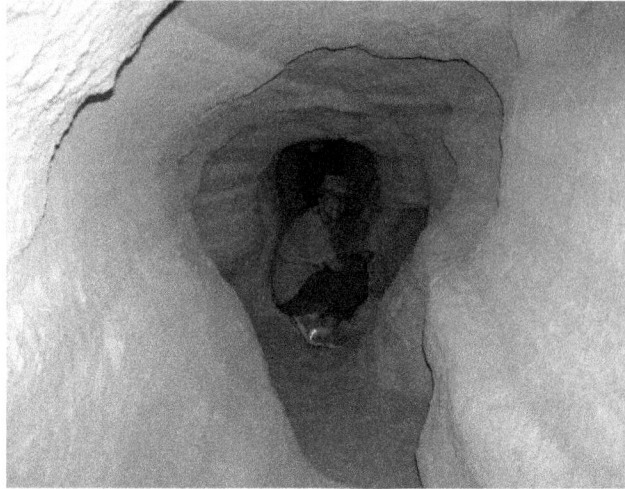

Chalcolithic copper mine in Timna Valley, Negev Desert

The oldest known mine on archaeological record is the "Lion Cave" in Swaziland, which radiocarbon dating shows to be about 43,000 years old. At this site Paleolithic humans mined hematite to make the red pigment ochre. Mines of a similar age in Hungary are believed to be sites where Neanderthals may have mined flint for weapons and tools.

## Ancient Egypt

Ancient Egyptians mined malachite at Maadi. At first, Egyptians used the bright green malachite stones for ornamentations and pottery. Later, between 2613 and 2494 BC, large building projects required expeditions abroad to the area of Wadi Maghareh in order to secure minerals and other resources not available in Egypt itself. Quarries for turquoise and copper were also found at Wadi Hamamat, Tura, Aswan and various other Nubian sites on the Sinai Peninsula and at Timna.

Mining in Egypt occurred in the earliest dynasties. The gold mines of Nubia were among the largest and most extensive of any in Ancient Egypt. These mines are described by the Greek author Diodorus Siculus, who mentions fire-setting as one method used to break down the hard rock holding the gold. One of the complexes is shown in one of the earliest known maps. The miners crushed the ore and ground it to a fine powder before washing the powder for the gold dust.

## Ancient Greek and Roman Mining

Mining in Europe has a very long history. Examples include the silver mines of Laurium, which helped support the Greek city state of Athens. Despite the mine having over 20,000 slaves working in them, the technology was essentially identical to their Bronze Age predecessors. Other mines, such as on the island of Thassos, had marble quarried by the Parians after having arrived in the 7th Century BC. The marble was shipped away and was later found by archaeologists to have been used in buildings including the tomb of Amphipolis. Philip II of Macedon, the father of Alexander the Great, captured the gold mines of Mount Pangeo in 357 BC to fund his military campaigns. He also captured gold mines in Thrace for minting coinage, eventually producing 26 tons per year.

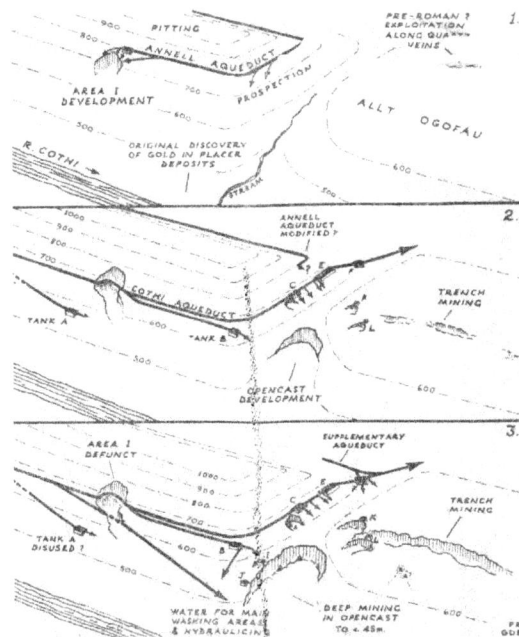

Ancient Roman development of the Dolaucothi Gold Mines, Wales

However, it was the Romans who developed large scale mining methods, especially the use of large volumes of water brought to the minehead by numerous aqueducts. The water was used for a variety of purposes, including removing overburden and rock debris, called hydraulic mining, as well as washing comminuted, or crushed, ores and driving simple machinery.

The Romans used hydraulic mining methods on a large scale to prospect for the veins of ore, especially a now obsolete form of mining known as hushing. This method involved building numerous aqueducts to supply water to the minehead where it was stored in large reservoirs and tanks. When a full tank was opened, the flood of water sluiced away the overburden to expose the bedrock underneath and any gold veins. The rock was then worked upon by fire-setting to heat the rock, which would be quenched with a stream of water. The resulting thermal shock cracked the rock, enabling it to be removed, aided by further streams of water from the overhead tanks. The Roman miners used similar methods to work cassiterite deposits in Cornwall and lead ore in the Pennines.

The methods had been developed by the Romans in Spain in 25 AD to exploit large alluvial gold deposits, the largest site being at Las Medulas, where seven long aqueducts were built to tap local rivers and to sluice the deposits. Spain was one of the most important mining regions, but all regions of the Roman Empire were exploited. In Great Britain the natives had mined minerals for millennia, but after the Roman conquest, the scale of the operations increased dramatically, as the Romans needed Britannia's resources, especially gold, silver, tin, and lead.

Roman techniques were not limited to surface mining. They followed the ore veins underground once opencast mining was no longer feasible. At Dolaucothi they stoped out the veins, and drove adits through barren rock to drain the stopes. The same adits were also used to ventilate the workings, especially important when fire-setting was used. At other parts of the site, they penetrated the water table and dewatered the mines using several kinds of machines, especially reverse overshot water-wheels. These were used extensively in the copper mines at Rio Tinto in Spain, where one sequence comprised 16 such wheels arranged in pairs, and lifting water about 80 feet (24 m). They were worked as treadmills with miners standing on the top slats. Many examples of such devices have been found in old Roman mines and some examples are now preserved in the British Museum and the National Museum of Wales.

## Medieval Europe

Gallery, 12th to 13th century, Germany

Mining as an industry underwent dramatic changes in medieval Europe. The mining industry in the early Middle Ages was mainly focused on the extraction of copper and iron. Other precious metals were also used, mainly for gilding or coinage. Initially, many metals were obtained through open-pit mining, and ore was primarily extracted from shallow depths, rather than through deep mine shafts. Around the 14th century, the growing use of weapons, armour, stirrups, and horseshoes greatly increased the demand for iron. Medieval knights, for example, were often laden with up to 100 pounds of plate or chain link armour in addition to swords, lances and other weapons. The overwhelming dependency on iron for military purposes spurred iron production and extraction processes.

The silver crisis of 1465 occurred when all mines had reached depths at which the shafts could no longer be pumped dry with the available technology. Although an increased use of bank notes, credit and copper coins during this period did decrease the value of, and dependence on, precious metals, gold and silver still remained vital to the story of medieval mining.

Due to differences in the social structure of society, the increasing extraction of mineral deposits spread from central Europe to England in the mid-sixteenth century. On the continent, all mineral deposits belonged to the crown, and this regalian right was stoutly maintained; but in England, it was pared down to gold and silver (of which there were virtually no deposits) by a judicial decision of 1568 and a law of 1688. England had iron, zinc, copper, lead, and tin ores. Landlords who owned the base metals and coal under their estates were now rendered with a strong inducement to extract these metals or to lease the deposits and collect royalties from mine operators. English,

German, and Dutch capital combined to finance extraction and refining. Hundreds of German technicians and skilled workers were brought over; in 1642 a colony of 4,000 foreigners was mining and smelting copper at Keswick in the northwestern mountains.

Use of water power in the form of water mills was extensive. The water mills were employed in crushing ore, raising ore from shafts, and ventilating galleries by powering giant bellows. Black powder was first used in mining in Selmecbánya, Kingdom of Hungary (now Banská Štiavnica, Slovakia) in 1627. Black powder allowed blasting of rock and earth to loosen and reveal ore veins. Blasting was much faster than fire-setting and allowed the mining of previously impenetrable metals and ores. In 1762, the world's first mining academy was established in the same town.

The widespread adoption of agricultural innovations such as the iron plowshare, as well as the growing use of metal as a building material, was also a driving force in the tremendous growth of the iron industry during this period. Inventions like the arrastra were often used by the Spanish to pulverize ore after being mined. This device was powered by animals and used the same principles used for grain threshing.

Much of the knowledge of medieval mining techniques comes from books such as Biringuccio's *De la pirotechnia* and probably most importantly from Georg Agricola's *De re metallica* (1556). These books detail many different mining methods used in German and Saxon mines. One of the prime issues confronting medieval miners (and one which Agricola explains in detail) was the removal of water from mining shafts. As miners dug deeper to access new veins, flooding became a very real obstacle. The mining industry became dramatically more efficient and prosperous with the invention of mechanical and animal driven pumps.

## Classical Philippine Civilization

The image of a Maharlika class of the Philippine Society , depicted in Boxer Codex that the Gold used as a form of Jewelry (ca.1400).

Mining in the Philippines began around 1000 BC. The early Filipinos worked various mines of gold, silver, copper and iron. Jewels, gold ingots, chains, calombigas and earrings were handed down from antiquity and inherited from their ancestors. Gold dagger handles, gold dishes, tooth plating, and huge gold ornamets were also used. In Laszlo Legeza's "Tantric elements in pre-Hispanic Philippines Gold Art", he mentioned that gold jewelry of Philippine origin was found in Ancient Egypt. According to Antonio Pigafetta, the people of Mindoro possessed great skill in mixing gold with other metals and gave it a natural and perfect appearance that could deceive even the best of silversmiths. The natives were also known for the jewelries made of other precious stones such as carnelian, agate and pearl. Some outstanding examples of Philippine jewelry included necklaces, belts, armlets and rings placed around the waist.

## The Americas

Lead mining in the upper Mississippi River region of the U.S., 1865.

There are ancient, prehistoric copper mines along Lake Superior, and metallic copper was still found there, near the surface, in colonial times.    Indegenous peoples availed themselves of this copper starting at least 5,000 years ago," and copper tools, arrowheads, and other artifacts that were part of an extensive native trade network have been discovered. In addition, obsidian, flint, and other minerals were mined, worked, and traded. Early French explorers who encountered the sites[clarification needed] made no use of the metals due to the difficulties of transporting them, but the copper was eventually traded throughout the continent along major river routes.

Miners at the Tamarack Mine in Copper Country, Michigan, U.S. in 1905.

In the early colonial history of the Americas, "native gold and silver was quickly expropriated and sent back to Spain in fleets of gold- and silver-laden galleons," the gold and silver originating mostly from mines in Central and South America. Turquoise dated at 700 A.D. was mined in pre-Columbian America; in the Cerillos Mining District in New Mexico, estimates are that "about 15,000 tons of rock had been removed from Mt. Chalchihuitl using stone tools before 1700."

Mining in the United States became prevalent in the 19th century, and the General Mining Act of 1872 was passed to encourage mining of federal lands. As with the California Gold Rush in the mid-19th century, mining for minerals and precious metals, along with ranching, was a driving factor in the Westward Expansion to the Pacific coast. With the exploration of the West, mining camps were established and "expressed a distinctive spirit, an enduring legacy to the new nation;" Gold Rushers would experience the same problems as the Land Rushers of the transient West that preceded them. Aided by railroads, many traveled West for work opportunities in mining. Western cities such as Denver and Sacramento originated as mining towns.

When new areas were explored, it was usually the gold (placer and then load) and then silver that were taken into possession and extracted first. Other metals would often wait for railroads or canals, as coarse gold dust and nuggets do not require smelting and are easy to identify and transport.

## Modern Period

In the early 20th century, the gold and silver rush to the western United States also stimulated mining for coal as well as base metals such as copper, lead, and iron. Areas in modern Montana, Utah, Arizona, and later Alaska became predominate suppliers of copper to the world, which was increasingly demanding copper for electrical and households goods. Canada's mining industry grew more slowly than did the United States' due to limitations in transportation, capital, and U.S. competition; Ontario was the major producer of the early 20th century with nickel, copper, and gold.

Meanwhile, Australia experienced the Australian gold rushes and by the 1850s was producing 40% of the world's gold, followed by the establishment of large mines such as the Mount Morgan Mine, which ran for nearly a hundred years, Broken Hill ore deposit (one of the largest zinc-lead ore deposits), and the iron ore mines at Iron Knob. After declines in production, another boom in mining occurred in the 1960s. Now, in the early 21st century, Australia remains a major world mineral producer.

As the 21st century begins, a globalized mining industry of large multinational corporations has arisen. Peak minerals and environmental impacts have also become a concern. Different elements, particularly rare earth minerals, have begun to increase in demand as a result of new technologies.

## Mine Development and Lifecycle

The process of mining from discovery of an ore body through extraction of minerals and finally to returning the land to its natural state consists of several distinct steps. The first is discovery of the ore body, which is carried out through prospecting or exploration to find and then define the extent, location and value of the ore body. This leads to a mathematical resource estimation to estimate the size and grade of the deposit.

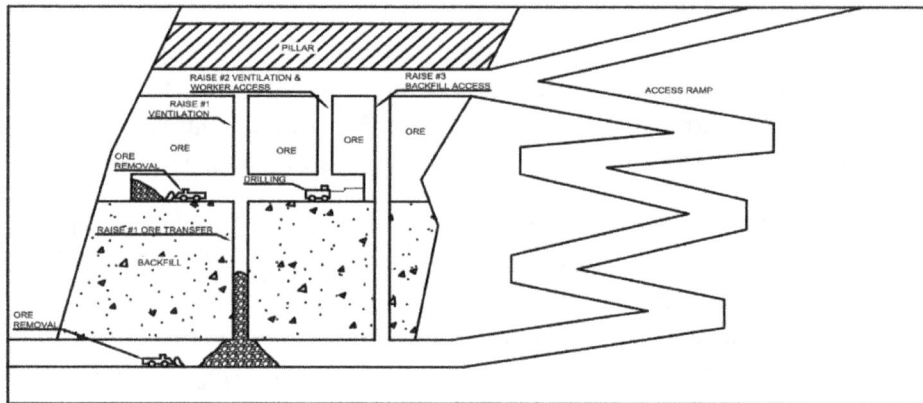

Schematic of a cut and fill mining operation in hard rock.

This estimation is used to conduct a pre-feasibility study to determine the theoretical economics of the ore deposit. This identifies, early on, whether further investment in estimation and engineering studies is warranted and identifies key risks and areas for further work. The next step is to conduct a feasibility study to evaluate the financial viability, the technical and financial risks, and the robustness of the project.

This is when the mining company makes the decision whether to develop the mine or to walk away from the project. This includes mine planning to evaluate the economically recoverable portion of the deposit, the metallurgy and ore recoverability, marketability and payability of the ore concentrates, engineering concerns, milling and infrastructure costs, finance and equity requirements, and an analysis of the proposed mine from the initial excavation all the way through to reclamation. The proportion of a deposit that is economically recoverable is dependent on the enrichment factor of the ore in the area.

To gain access to the mineral deposit within an area it is often necessary to mine through or remove waste material which is not of immediate interest to the miner. The total movement of ore and waste constitutes the mining process. Often more waste than ore is mined during the life of a mine, depending on the nature and location of the ore body. Waste removal and placement is a major cost to the mining operator, so a detailed characterization of the waste material forms an essential part of the geological exploration program for a mining operation.

Once the analysis determines a given ore body is worth recovering, development begins to create access to the ore body. The mine buildings and processing plants are built, and any necessary equipment is obtained. The operation of the mine to recover the ore begins and continues as long as the company operating the mine finds it economical to do so. Once all the ore that the mine can produce profitably is recovered, reclamation begins to make the land used by the mine suitable for future use.

## Mining Techniques

Mining techniques can be divided into two common excavation types: surface mining and sub-surface (underground) mining. Today, surface mining is much more common, and produces, for example, 85% of minerals (excluding petroleum and natural gas) in the United States, including 98% of metallic ores.

Underground longwall mining.

Targets are divided into two general categories of materials: *placer deposits*, consisting of valuable minerals contained within river gravels, beach sands, and other unconsolidated materials; and *lode deposits*, where valuable minerals are found in veins, in layers, or in mineral grains generally distributed throughout a mass of actual rock. Both types of ore deposit, placer or lode, are mined by both surface and underground methods.

Some mining, including much of the rare earth elements and uranium mining, is done by less-common methods, such as in-situ leaching: this technique involves digging neither at the surface nor underground. The extraction of target minerals by this technique requires that they be soluble, e.g., potash, potassium chloride, sodium chloride, sodium sulfate, which dissolve in water. Some minerals, such as copper minerals and uranium oxide, require acid or carbonate solutions to dissolve.

## Surface Mining

Surface mining is done by removing (stripping) surface vegetation, dirt, and, if necessary, layers of bedrock in order to reach buried ore deposits. Techniques of surface mining include: open-pit mining, which is the recovery of materials from an open pit in the ground, quarrying, identical to open-pit mining except that it refers to sand, stone and clay; strip mining, which consists of stripping surface layers off to reveal ore/seams underneath; and mountaintop removal, commonly associated with coal mining, which involves taking the top of a mountain off to reach ore deposits at depth. Most (but not all) placer deposits, because of their shallowly buried nature, are mined by surface methods. Finally, landfill mining involves sites where landfills are excavated and processed.

Garzweiler surface mine, Germany

## Underground Mining

Mantrip used for transporting miners within an underground mine

Sub-surface mining consists of digging tunnels or shafts into the earth to reach buried ore deposits. Ore, for processing, and waste rock, for disposal, are brought to the surface through the tunnels and shafts. Sub-surface mining can be classified by the type of access shafts used, the extraction method or the technique used to reach the mineral deposit. Drift mining utilizes horizontal access tunnels, slope mining uses diagonally sloping access shafts, and shaft mining utilizes vertical access shafts. Mining in hard and soft rock formations require different techniques.

Other methods include shrinkage stope mining, which is mining upward, creating a sloping underground room, long wall mining, which is grinding a long ore surface underground, and room and pillar mining, which is removing ore from rooms while leaving pillars in place to support the roof of the room. Room and pillar mining often leads to retreat mining, in which supporting pillars are removed as miners retreat, allowing the room to cave in, thereby loosening more ore. Additional sub-surface mining methods include hard rock mining, which is mining of hard rock (igneous, metamorphic or sedimentary) materials, bore hole mining, drift and fill mining, long hole slope mining, sub level caving, and block caving.

## Highwall Mining

Caterpillar Highwall Miner HW300 - Technology Bridging Underground and Open Pit Mining

Highwall mining is another form of surface mining that evolved from auger mining. In Highwall mining, the coal seam is penetrated by a continuous miner propelled by a hydraulic Pushbeam Transfer Mechanism (PTM). A typical cycle includes sumping (launch-pushing forward) and shearing (raising and lowering the cutterhead boom to cut the entire height of the coal seam). As the coal recovery cycle continues, the cutterhead is progressively launched into the coal seam for 19.72 feet (6.01 m). Then, the Pushbeam Transfer Mechanism (PTM) automatically inserts a 19.72-foot (6.01 m) long rectangular Pushbeam (Screw-Conveyor Segment) into the center section of the machine between the Powerhead and the cutterhead. The Pushbeam system can penetrate nearly 1,000 feet (300 m) into the coal seam. One patented Highwall mining systems use augers enclosed inside the Pushbeam that prevent the mined coal from being contaminated by rock debris during the conveyance process. Using a video imaging and/or a gamma ray sensor and/or other Geo-Radar systems like a coal-rock interface detection sensor (CID), the operator can see ahead projection of the seam-rock interface and guide the continuous miner's progress. Highwall mining can produce thousands of tons of coal in contour-strip operations with narrow benches, previously mined areas, trench mine applications and steep-dip seams with controlled water-inflow pump system and/or a gas (inert) venting system.

## Machines

Heavy machinery is used in mining to explore and develop sites, to remove and stockpile overburden, to break and remove rocks of various hardness and toughness, to process the ore, and to carry out reclamation projects after the mine is closed. Bulldozers, drills, explosives and trucks are all necessary for excavating the land. In the case of placer mining, unconsolidated gravel, or alluvium, is fed into machinery consisting of a hopper and a shaking screen or trommel which frees the desired minerals from the waste gravel. The minerals are then concentrated using sluices or jigs.

The Bagger 288 is a bucket-wheel excavator used in strip mining. It is also the largest land vehicle of all time.

A Bucyrus Erie 2570 dragline and CAT 797 haul truck at the North Antelope Rochelle opencut coal mine

Large drills are used to sink shafts, excavate stopes, and obtain samples for analysis. Trams are used to transport miners, minerals and waste. Lifts carry miners into and out of mines, and move rock and ore out, and machinery in and out, of underground mines. Huge trucks, shovels and cranes are employed in surface mining to move large quantities of overburden and ore. Processing plants utilize large crushers, mills, reactors, roasters and other equipment to consolidate the mineral-rich material and extract the desired compounds and metals from the ore.

## Processing

Once the mineral is extracted, it is often then processed. The science of extractive metallurgy is a specialized area in the science of metallurgy that studies the extraction of valuable metals from their ores, especially through chemical or mechanical means.

Mineral processing (or mineral dressing) is a specialized area in the science of metallurgy that studies the mechanical means of crushing, grinding, and washing that enable the separation (extractive metallurgy) of valuable metals or minerals from their gangue (waste material). Processing of placer ore material consists of gravity-dependent methods of separation, such as sluice boxes. Only minor shaking or washing may be necessary to disaggregate (unclump) the sands or gravels before processing. Processing of ore from a lode mine, whether it is a surface or subsurface mine, requires that the rock ore be crushed and pulverized before extraction of the valuable minerals begins. After lode ore is crushed, recovery of the valuable minerals is done by one, or a combination of several, mechanical and chemical techniques.

Since most metals are present in ores as oxides or sulfides, the metal needs to be reduced to its metallic form. This can be accomplished through chemical means such as smelting or through electrolytic reduction, as in the case of aluminium. Geometallurgy combines the geologic sciences with extractive metallurgy and mining.

## Environmental Effects

Iron hydroxide precipitate stains a stream receiving acid drainage from surface coal mining.

Environmental issues can include erosion, formation of sinkholes, loss of biodiversity, and contamination of soil, groundwater and surface water by chemicals from mining processes. In some cases, additional forest logging is done in the vicinity of mines to create space for the storage of the created debris and soil. Contamination resulting from leakage of chemicals can also affect the health of the local population if not properly controlled. Extreme examples of pollution from mining activities include coal fires, which can last for years or even decades, producing massive amounts of environmental damage.

Mining companies in most countries are required to follow stringent environmental and rehabilitation codes in order to minimize environmental impact and avoid impacting human health. These codes and regulations all require the common steps of environmental impact assessment, development of environmental management plans, mine closure planning (which must be done before the start of mining operations), and environmental monitoring during operation and after closure. However, in some areas, particularly in the developing world, government regulations may not be well enforced.

For major mining companies and any company seeking international financing, there are a number of other mechanisms to enforce good environmental standards. These generally relate to financing standards such as the Equator Principles, IFC environmental standards, and criteria for Socially responsible investing. Mining companies have used this oversight from the financial sector to argue for some level of industry self-regulation. In 1992, a Draft Code of Conduct for Transnational Corporations was proposed at the Rio Earth Summit by the UN Centre for Transnational Corporations (UNCTC), but the Business Council for Sustainable Development (BCSD) together with the International Chamber of Commerce (ICC) argued successfully for self-regulation instead.

This was followed by the Global Mining Initiative which was begun by nine of the largest metals and mining companies and which led to the formation of the International Council on Mining and Metals, whose purpose was to "act as a catalyst" in an effort to improve social and environmental performance in the mining and metals industry internationally. The mining industry has provided funding to various conservation groups, some of which have been working with conservation agendas that are at odds with an emerging acceptance of the rights of indigenous people – particularly the right to make land-use decisions.

Certification of mines with good practices occurs through the International Organization for Standardization (ISO). For example, ISO 9000 and ISO 14001, which certify an "auditable environmental management system", involve short inspections, although they have been accused of lacking rigor. Certification is also available through Ceres' Global Reporting Initiative, but these reports are voluntary and unverified. Miscellaneous other certification programs exist for various projects, typically through nonprofit groups.

The purpose of a 2012 EPS PEAKS paper was to provide evidence on policies managing ecological costs and maximise socio-economic benefits of mining using host country regulatory initiatives. It found existing literature suggesting donors encourage developing countries to:

- Make the environment-poverty link and introduce cutting-edge wealth measures and natural capital accounts.

- Reform old taxes in line with more recent financial innovation, engage directly with the companies, enacting land use and impact assessments, and incorporate specialised support and standards agencies.

- Set in play transparency and community participation initiatives using the wealth accrued.

## Waste

Ore mills generate large amounts of waste, called tailings. For example, 99 tons of waste are generated per ton of copper, with even higher ratios in gold mining - because only 5.3 g of gold is extracted per ton of ore, a ton of gold produces 200,000 tons of tailings. These tailings can be toxic. Tailings, which are usually produced as a slurry, are most commonly dumped into ponds made from naturally existing valleys. These ponds are secured by impoundments (dams or embankment dams). In 2000 it was estimated that 3,500 tailings impoundments existed, and that every year, 2 to 5 major failures and 35 minor failures occurred; for example, in the Marcopper mining disaster at least 2 million tons of tailings were released into a local river. Subaqueous tailings disposal is another option. The mining industry has argued that submarine tailings disposal (STD), which disposes of tailings in the sea, is ideal because it avoids the risks of tailings ponds; although the practice is illegal in the United States and Canada, it is used in the developing world.

The waste is classified as either sterile or mineralised, with acid generating potential, and the movement and storage of this material forms a major part of the mine planning process. When the mineralised package is determined by an economic cut-off, the near-grade mineralised waste is usually dumped separately with view to later treatment should market conditions change and it becomes economically viable. Civil engineering design parameters are used in the design of the waste dumps, and special conditions apply to high-rainfall areas and to seismically active areas. Waste dump designs must meet all regulatory requirements of the country in whose jurisdiction the mine is located. It is also common practice to rehabilitate dumps to an internationally acceptable standard, which in some cases means that higher standards than the local regulatory standard are applied.

## Renewable Energy and Mining

Many mining sites are remote and not connected to the grid. Electricity is typically generated with diesel generators. Due to high transportation cost and theft during transportation the cost for generating electricity is normally high. Renewable energy applications are becoming an alternative or amendment. Both solar and wind power plants can contribute in saving diesel costs at mining sites. Renewable energy applications have been built at mining sites. Cost savings can reach up to 70%.

## Mining Industry

Mining exists in many countries. London is known as the capital of global "mining houses" such as Rio Tinto Group, BHP Billiton, and Anglo American PLC. The US mining industry is also large, but it is dominated by the coal and other nonmetal minerals (e.g., rock and sand), and various regulations have worked to reduce the significance of mining in the United States. In 2007 the total market capitalization of mining companies was reported at US$962 billion, which compares to a total global market cap of publicly traded companies of about US$50 trillion in 2007. In 2002, Chile and Peru were reportedly the major mining countries of South America. The mineral industry of Africa includes the mining of various minerals; it produces relatively little of the industrial metals copper, lead, and zinc, but according to one estimate has as a percent of world reserves 40% of gold, 60% of cobalt, and 90% of the world's platinum group metals. Mining in India is a significant part of that country's economy. In the developed world, mining in Australia, with BHP Billiton founded and headquartered in the country, and mining in Canada are particularly significant. For rare earth minerals mining, China reportedly controlled 95% of production in 2013.

The Bingham Canyon Mine of Rio Tinto's subsidiary, Kennecott Utah Copper.

While exploration and mining can be conducted by individual entrepreneurs or small businesses, most modern-day mines are large enterprises requiring large amounts of capital to establish. Consequently, the mining sector of the industry is dominated by large, often multinational, companies, most of them publicly listed. It can be argued that what is referred to as the 'mining industry' is actually two sectors, one specializing in exploration for new resources and the other in mining those resources. The exploration sector is typically made up of individuals and small mineral resource companies, called "juniors", which are dependent on venture capital. The mining sector is made up of large multinational companies that are sustained by production from their mining operations. Various other industries such as equipment manufacture, environmental testing, and metallurgy analysis rely on, and support, the mining industry throughout the world. Canadian stock exchanges have a particular focus on mining companies, particularly junior exploration companies through Toronto's TSX Venture Exchange; Canadian companies raise capital on these exchanges and then invest the money in exploration globally. Some have argued that below juniors there exists a substantial sector of illegitimate companies primarily focused on manipulating stock prices.

Mining operations can be grouped into five major categories in terms of their respective resources. These are oil and gas extraction, coal mining, metal ore mining, nonmetallic mineral mining and quarrying, and mining support activities. Of all of these categories, oil and gas extraction remains one of the largest in terms of its global economic importance. Prospecting potential mining sites, a vital area of concern for the mining industry, is now done using sophisticated new technologies such as seismic prospecting and remote-sensing satellites. Mining is heavily affected by the prices of the commodity minerals, which are often volatile. The 2000s commodities boom ("commodities supercycle") increased the prices of commodities, driving aggressive mining. In addition, the price of gold increased dramatically in the 2000s, which increased gold mining; for example, one study found that conversion of forest in the Amazon increased six-fold from the period 2003–2006 (292 ha/yr) to the period 2006–2009 (1,915 ha/yr), largely due to artisanal mining.

## Corporate Classifications

Mining companies can be classified based on their size and financial capabilities:

- Major companies are considered to have an adjusted annual mining-related revenue of more than US$500 million, with the financial capability to develop a major mine on its own.

- Intermediate companies have at least $50 million in annual revenue but less than $500 million.

- Junior companies rely on equity financing as their principal means of funding exploration. Juniors are mainly pure exploration companies, but may also produce minimally, and do not have a revenue exceeding US$50 million.

## Regulation and Governance

New regulation and process of legislative reforms aims to enrich the harmonization and stability of the mining sector in mineral-rich countries. The new legislation for mining industry in the African countries still appears as an emerging issue with a potential to be solved, until a consensus is reached on the best approach. By the beginning of the 21st century the booming and more complex mining sector in mineral-rich countries provided only slight benefits to local communities in terms of sustainability. Increasing debates and influence by NGOs and communities appealed for a new program which would have had also included a disadvantaged communities, and would have had worked towards sustainable development even after mine closure (included transparency and revenue management). By the early 2000s, community development issues and resettlements became mainstreamed in World Bank mining projects. Mining-industry expansion after an increase of mineral prices in 2003 and also potential fiscal revenues in those countries created an omission in the other economic sectors in terms of finances and development. Furthermore, it had highlighted regional and local demand of mining-revenues and lack of ability of sub-national governments to use the revenues. The Fraser Institute (a Canadian think tank) has highlighted the environmental protection laws in developing countries, as well as the voluntary efforts by mining companies to improve their environmental impact.

In 2007 the Extractive Industries Transparency Initiative (EITI) was mainstreamed in all countries cooperating with the World Bank in mining industry reform. The EITI is operating and implementing with a support of EITI Multi-Donor Trust Fund, managed by The World Bank. The Extractive Industries Transparency Initiative (EITI) aims to increase transparency in transactions between governments and companies within extractive industries by monitoring the revenues and benefits between industries and recipient governments. The entrance process is voluntary for each country and is being monitored by multi-stakeholders involving government, private companies and civil society representatives, responsible for disclosure and dissemination of the reconciliation report; however, the competitive disadvantage of company-by company public report is for some of the businesses in Ghana, the main constraint. Therefore, the outcome assessment in terms of failure or success of the new EITI regulation does not only "rest on the government's shoulders" but also on civil society and companies.

On the other hand, criticism points out two main implementation issues; inclusion or exclusion of artisanal mining and small-scale mining (ASM) from the EITI and how to deal with "non-cash" payments made by companies to subnational governments. Furthermore, disproportion of the revenues mining industry creates to the comparatively small number of people that it employs, causes another controversy. The issue of artisanal mining is clearly an issue in EITI Countries such as the Central African Republic, D.R. Congo, Guinea, Liberia and Sierra Leone – i.e. almost half of the mining countries implementing the EITI. Among other things, limited scope of the EITI involving disparity in terms of knowledge of the industry and negotiation skills, thus far flexibility of the policy (e.g. liberty of the countries to expand beyond the minimum requirements and adapt it to their needs), creates another risk of unsuccessful implementation. Public awareness increase, where government should act as a bridge between public and initiative for a successful outcome of the policy is an important element to be considered.

## World Bank

The World Bank has been involved in mining since 1955, mainly through grants from its International Bank for Reconstruction and Development, with the Bank's Multilateral Investment Guarantee Agency offering political risk insurance. Between 1955 and 1990 it provided about $2 billion to fifty mining projects, broadly categorized as reform and rehabilitation, greenfield mine construction, mineral processing, technical assistance, and engineering. These projects have been criticized, particularly the Ferro Carajas project of Brazil, begun in 1981. The World Bank established mining codes intended to increase foreign investment; in 1988 it solicited feedback from 45 mining companies on how to increase their involvement.

In 1992 the World Bank began to push for privatization of government-owned mining companies with a new set of codes, beginning with its report *The Strategy for African Mining*. In 1997, Latin America's largest miner Companhia Vale do Rio Doce (CVRD) was privatized. These and other developments such as the Philippines 1995 Mining Act led the bank to publish a third report (*Assistance for Minerals Sector Development and Reform in Member Countries*) which endorsed mandatory environment impact assessments and attention to the concerns of the local population. The codes based on this report are influential in the legislation of developing nations. The new codes are intended to encourage development through tax holidays, zero custom duties, reduced income taxes, and related measures. The results of these codes were analyzed by a group from the University of Quebec, which concluded that the codes promote foreign investment but "fall very short of permitting sustainable development". The observed negative correlation between natural resources and economic development is known as the resource curse.

## Safety

Safety has long been a concern in the mining business, especially in sub-surface mining. The Courrières mine disaster, Europe's worst mining accident, involved the death of 1,099 miners in Northern France on March 10, 1906. This disaster was surpassed only by the Benxihu Colliery accident in China on April 26, 1942, which killed 1,549 miners. While mining today is substantially safer than it was in previous decades, mining accidents still occur. Government figures indicate that

5,000 Chinese miners die in accidents each year, while other reports have suggested a figure as high as 20,000. Mining accidents continue worldwide, including accidents causing dozens of fatalities at a time such as the 2007 Ulyanovskaya Mine disaster in Russia, the 2009 Heilongjiang mine explosion in China, and the 2010 Upper Big Branch Mine disaster in the United States.

There are numerous occupational hazards associated with mining, including exposure to rockdust which can lead to diseases such as silicosis, asbestosis, and pneumoconiosis. Gases in the mine can lead to asphyxiation and could also be ignited. Mining equipment can generate considerable noise, putting workers at risk for hearing loss. Cave-ins, rock falls, and exposure to excess heat are also known hazards.

Proper ventilation, hearing protection, and spraying equipment with water are important safety practices in mines.

## Records

As of 2008, the deepest mine in the world is TauTona in Carletonville, South Africa at 3.9 kilometres (2.4 mi), replacing the neighboring Savuka Mine in the North West Province of South Africa at 3,774 metres (12,382 ft). East Rand Mine in Boksburg, South Africa briefly held the record at 3,585 metres (11,762 ft), and the first mine declared the deepest in the world was also TauTona when it was at 3,581 metres (11,749 ft).

The Moab Khutsong gold mine in North West Province (South Africa) has the world's longest winding steel wire rope, able to lower workers to 3,054 metres (10,020 ft) in one uninterrupted four-minute journey.

The deepest mine in Europe is the 16th shaft of the uranium mines in Příbram, Czech Republic at 1,838 metres (6,030 ft), second is Bergwerk Saar in Saarland, Germany at 1,750 metres (5,740 ft).

The deepest open-pit mine in the world is Bingham Canyon Mine in Bingham Canyon, Utah, United States at over 1,200 metres (3,900 ft). The largest and second deepest open-pit copper mine in the world is Chuquicamata in Chuquicamata, Chile at 900 metres (3,000 ft), 443,000 tons of copper and 20,000 tons of molybdenum produced annually.

The deepest open-pit mine with respect to sea level is Tagebau Hambach in Germany, where the base of the pit is 293 metres (961 ft) below sea level.

The largest underground mine is Kiirunavaara Mine in Kiruna, Sweden. With 450 kilometres (280 mi) of roads, 40 million tonnes of ore produced yearly, and a depth of 1,270 metres (4,170 ft), it is also one of the most modern underground mines. The deepest borehole in the world is Kola Superdeep Borehole at 12,262 metres (40,230 ft). This, however, is not a matter of mining but rather related to scientific drilling.

## Metal Reserves and Recycling

During the twentieth century, the variety of metals used in society grew rapidly. Today, the development of major nations such as China and India and advances in technologies are fueling an ever greater demand. The result is that metal mining activities are expanding and more and more of the world's metal stocks are above ground in use rather than below ground as unused reserves. An example is the in-use stock of copper. Between 1932 and 1999, copper in use in the USA rose from 73 kilograms (161 lb) to 238 kilograms (525 lb) per person.

95% of the energy used to make aluminium from bauxite ore is saved by using recycled material. However, levels of metals recycling are generally low. In 2010, the International Resource Panel, hosted by the United Nations Environment Programme (UNEP), published reports on metal stocks that exist within society and their recycling rates.

The report's authors observed that the metal stocks in society can serve as huge mines above ground. However, they warned that the recycling rates of some rare metals used in applications such as mobile phones, battery packs for hybrid cars, and fuel cells are so low that unless future end-of-life recycling rates are dramatically stepped up these critical metals will become unavailable for use in modern technology.

As recycling rates are low and so much metal has already been extracted, some landfills now contain higher concentrations of metal than mines themselves. This is especially true with aluminium, found in cans, and precious metals in discarded electronics. Furthermore, waste after 15 years has still not broken down, so less processing would be required when compared to mining ores. A study undertaken by Cranfield University has found £360 million of metals could be mined from just 4 landfill sites. There is also up to 20MW/kg of energy in waste, potentially making the re-extraction more profitable. However, although the first landfill mine opened in Tel Aviv, Israel in 1953, little work has followed due to the abundance of accessible ores.

# Casting

Casting is a manufacturing process in which a liquid material is usually poured into a mold, which contains a hollow cavity of the desired shape, and then allowed to solidify. The solidified part is also known as a casting, which is ejected or broken out of the mold to complete the process. Casting materials are usually metals or various cold setting materials that cure after mixing two or more components together; examples are epoxy, concrete, plaster and clay. Casting is most often used for making complex shapes that would be otherwise difficult or uneconomical to make by other methods.

Casting is a 6000-year-old process. The oldest surviving casting is a copper frog from 3200 BC.

Molten metal before casting

Casting iron in a sand mold

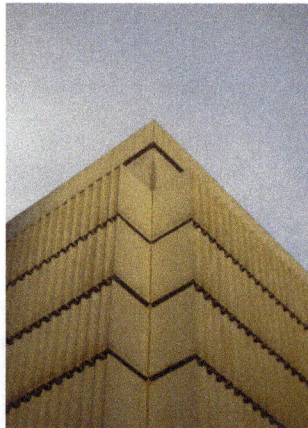

Judenplatz Holocaust Memorial (Nameless Library), by Rachel Whiteread. Concrete cast of books on library shelves turned inside out.

## Types

## Metal

In metalworking, metal is heated until it becomes liquid and is then poured into a mold. The mold is a hollow cavity that includes the desired shape, but the mold also includes runners and vents that enable the metal to fill the mold. The mold and the metal are then cooled until the metal solidifies. The solidified part (the casting) is then recovered from the mold. Subsequent operations remove excess material caused by the casting process (e.g., such as the runners and vents).

## Plaster, Concrete, or Plastic Resin

Plaster and other chemical curing materials such as concrete and plastic resin may be cast using single-use *waste* molds as noted above, multiple-use 'piece' molds, or molds made of small rigid pieces or of flexible material such as latex rubber (which is in turn supported by an exterior mold). When casting plaster or concrete, the material surface is flat and lacks transparency. Often topical treatments are applied to the surface. For example, painting and etching can be used in a way that give the appearance of metal or stone. Alternatively, the material is altered in its initial casting process and may contain colored sand so as to give an appearance of stone. By casting concrete, rather than plaster, it is possible to create sculptures, fountains, or seating for outdoor use. A simulation of high-quality marble may be made using certain chemically-set plastic resins (for example epoxy or polyester) with powdered stone added for coloration, often with multiple colors worked in. The latter is a common means of making washstands, washstand tops and shower stalls, with the skilled working of multiple colors resulting in simulated staining patterns as is often found in natural marble or travertine.

## Fettling

Raw castings often contain irregularities caused by seams and imperfections in the molds, as well as access ports for pouring material into the molds. The process of cutting, grinding, shaving or sanding away these unwanted bits is called "fettling". In modern times robotic processes have been developed to perform some of the more repetitive parts of the fettling process, but historically fettlers carried out this arduous work manually, and often in conditions dangerous to their health.

Fettling can add significantly to the cost of the resulting product, and designers of molds seek to minimize it through the shape of the mold, the material being cast, and sometimes by including decorative elements.

## Casting Process Simulation

Casting process simulation uses numerical methods to calculate cast component quality considering mold filling, solidification and cooling, and provides a quantitative prediction of casting mechanical properties, thermal stresses and distortion. Simulation accurately describes a cast component's quality up-front before production starts. The casting rigging can be designed with respect to the required component properties. This has benefits beyond a reduction in pre-production sampling, as the precise layout of the complete casting system also leads to energy, material, and tooling savings.

The software supports the user in component design, the determination of melting practice and casting methoding through to pattern and mold making, heat treatment, and finishing. This saves costs along the entire casting manufacturing route.

Casting process simulation was initially developed at universities starting from the early '70s, mainly in Europe and in the U.S., and is regarded as the most important innovation in casting technology over the last 50 years. Since the late '80s, commercial programs (such as AutoCAST and MAGMA) are available which make it possible for foundries to gain new insight into what is happening inside the mold or die during the casting process.

## Melting

Melting (also known as fusion) is a physical process that results in the phase transition of a substance from a solid to a liquid. This occurs when the internal energy of the solid increases, typically by the application of heat or pressure, which increases the substance's temperature to the melting point. At the melting point, the ordering of ions or molecules in the solid breaks down to a less ordered state, and the solid melts to become a liquid. An object that has melted completely is molten (although this word is typically used for substances that melt only at a high temperature, such as molten iron or molten lava).

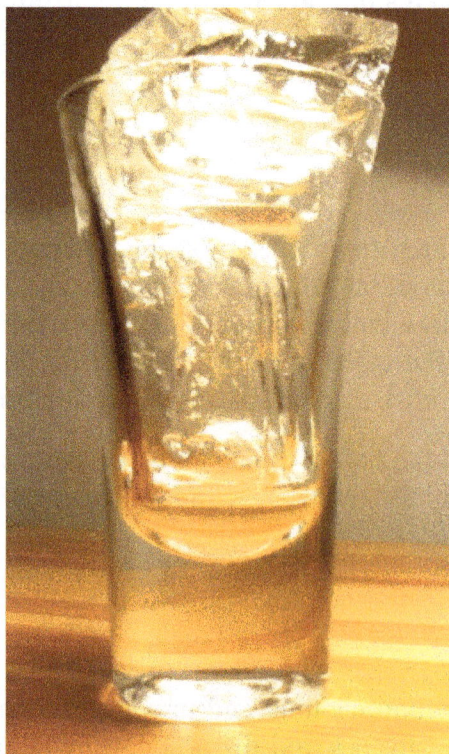

Ice melting

Substances in the molten state generally have reduced viscosity as the temperature increases. An exception to this principle is the element sulfur, whose viscosity increases to a point due to polymerization and then decreases with higher temperatures in its molten state.

Some organic compounds melt through mesophases, states of partial order between solid and liquid.

## Melting as a First-Order Phase Transition

From a thermodynamics point of view, at the melting point the change in Gibbs free energy $\Delta G$ of the material is zero, but there are non-zero changes in the enthalpy ($H$) and the entropy ($S$), known respectively as the enthalpy of fusion (or latent heat of fusion) and the entropy of fusion. Melting is therefore classified as a first-order phase transition. Melting occurs when the Gibbs free energy of the liquid becomes lower than the solid for that material. The temperature at which this occurs is dependent on the ambient pressure.

Low-temperature helium is the only known exception to the general rule. Helium-3 has a negative enthalpy of fusion at temperatures below 0.3 K. Helium-4 also has a very slightly negative enthalpy of fusion below 0.8 K. This means that, at appropriate constant pressures, heat must be *removed* from these substances in order to melt them.

## Melting Criteria

Among the theoretical criteria for melting, the Lindemann and of Born criteria are those most frequently used as a basis to analyse the melting conditions . The Lindemann criterion states that melting occurs because of vibrational instability, e.g. crystals melt when the average amplitude of thermal vibrations of atoms is relatively high compared with interatomic distances, e.g. $<\delta u^2>^{1/2} > \delta_L R_s$, where $\delta u$ is the atomic displacement, the Lindemann parameter $\delta_L \approx 0.20...0.25$ and $R_s$ is one-half of the inter-atomic distance. The Lindemann melting criterion is supported by experimental data both for crystalline materials and for glass-liquid transitions in amorphous materials. The Born criterion is based on rigidity catastrophe caused by the vanishing elastic shear modulus, e.g. when the crystal no longer has sufficient rigidity to mechanically withstand load.

## Supercooling

Under a standard set of conditions, the melting point of a substance is a characteristic property. The melting point is often equal to the freezing point. However, under carefully created conditions, supercooling or superheating past the melting or freezing point can occur. Water on a very clean glass surface will often supercool several degrees below the freezing point without freezing. Fine emulsions of pure water have been cooled to −38 degrees Celsius without nucleation to form ice.. Nucleation occurs due to fluctuations in the properties of the material. If the material is kept still there is often nothing (such a physical vibration) to trigger this change, and supercooling (or superheating) may occur. Thermodynamically, the supercooled liquid is in the metastable state with respect to the crystalline phase, and it is likely to crystallize suddenly.

## Melting of Amorphous Solids (Glasses)

Glasses are amorphous solids which are usually fabricated when the molten material cools very rapidly to below its glass transition temperature, without sufficient time for a regular crystal lattice to form. Solids are characterised by a high degree of connectivity between their molecules, and fluids have lower connectivity of their structural blocks. Melting of a solid material can also be considered as a percolation via broken connections between particles e.g. connecting bonds. In this approach melting of an amorphous material occurs when the broken bonds form a percolation cluster with $T_g$ dependent on quasi-equilibrium thermodynamic parameters of bonds e.g. on enthalpy ($H_d$) and entropy ($S_d$) of formation of bonds in a given system at given conditions:

$$T_g = \frac{H_d}{S_d + R\ln(\frac{1-f_c}{f_c})}$$

where $f_c$ is the percolation threshold and $R$ is the universal gas constant. Although $H_d$ and $S_d$ are not true equilibrium thermodynamic parameters and can depend on the cooling rate of a melt they can be found from available experimental data on viscosity of amorphous materials.

Even below its melting point, quasi-liquid films can be observed on crystalline surfaces. The thickness of the film is temperature dependent. This effect is common for all crystalline materials. Premelting shows its effects in e.g. frost heave, the growth of snowflakes and, taking grain boundary interfaces into account, maybe even in the movement of glaciers.

## Related Concepts

In genetics, melting DNA means to separate the double-stranded DNA into two single strands by heating or the use of chemical agents, cf. polymerase chain reaction.

# Welding

Gas metal arc welding (MIG welding)

**Welding** is a fabrication or sculptural process that joins materials, usually metals or thermoplastics, by causing fusion, which is distinct from lower temperature metal-joining techniques such as brazing and soldering, which do not melt the base metal. In addition to melting the base metal, a filler material is often added to the joint to form a pool of molten material (the weld pool) that cools to form a joint that can be as strong, or even stronger, than the base material. Pressure may also be used in conjunction with heat, or by itself, to produce a weld.

Although less common, there are also solid state welding processes such as friction welding or shielded active gas welding in which metal does not melt.

Some of the best known welding methods include:

- Shielded metal arc welding (SMAW) – also known as "stick welding or electric welding", uses an electrode that has flux around it to protect the weld puddle. The electrode holder holds the electrode as it slowly melts away. Slag protects the weld puddle from atmospheric contamination.

- Gas tungsten arc welding (GTAW) – also known as TIG (tungsten, inert gas), uses a non-consumable tungsten electrode to produce the weld. The weld area is protected from atmospheric contamination by an inert shielding gas such as argon or helium.

- Gas metal arc welding (GMAW) – commonly termed MIG (metal, inert gas), uses a wire feeding gun that feeds wire at an adjustable speed and flows an argon-based shielding gas or a mix of argon and carbon dioxide ($CO_2$) over the weld puddle to protect it from atmospheric contamination.

- Flux-cored arc welding (FCAW) – almost identical to MIG welding except it uses a special tubular wire filled with flux; it can be used with or without shielding gas, depending on the filler.

- Submerged arc welding (SAW) – uses an automatically fed consumable electrode and a blanket of granular fusible flux. The molten weld and the arc zone are protected from atmospheric contamination by being "submerged" under the flux blanket.

- Electroslag welding (ESW) – a highly productive, single pass welding process for thicker materials between 1 inch (25 mm) and 12 inches (300 mm) in a vertical or close to vertical position.

Many different energy sources can be used for welding, including a gas flame, an electric arc, a laser, an electron beam, friction, and ultrasound. While often an industrial process, welding may be performed in many different environments, including in open air, under water, and in outer space. Welding is a hazardous undertaking and precautions are required to avoid burns, electric shock, vision damage, inhalation of poisonous gases and fumes, and exposure to intense ultraviolet radiation.

Until the end of the 19th century, the only welding process was forge welding, which blacksmiths had used for centuries to join iron and steel by heating and hammering. Arc welding and oxyfuel welding were among the first processes to develop late in the century, and electric resistance welding followed soon after. Welding technology advanced quickly during the early 20th century as the world wars drove the demand for reliable and inexpensive joining methods. Following the wars, several modern welding techniques were developed, including manual methods like SMAW, now one of the most popular welding methods, as well as semi-automatic and automatic processes such as GMAW, SAW, FCAW and ESW. Developments continued with the invention of laser beam welding, electron beam welding, magnetic pulse welding (MPW), and friction stir welding in the latter half of the century. Today, the science continues to advance. Robot welding is commonplace in industrial settings, and researchers continue to develop new welding methods and gain greater understanding of weld quality.

## History

The history of joining metals goes back several millennia. Called forge welding, the earliest examples come from the Bronze and Iron Ages in Europe and the Middle East. The ancient Greek historian Herodotus states in *The Histories* of the 5th century BC that Glaucus of Chios "was the man who single-handedly invented iron welding". Welding was used in the construction of the Iron pillar of Delhi, erected in Delhi, India about 310 AD and weighing 5.4 metric tons.

The Middle Ages brought advances in forge welding, in which blacksmiths pounded heated metal repeatedly until bonding occurred. In 1540, Vannoccio Biringuccio published *De la pirotechnia*, which includes descriptions of the forging operation. Renaissance craftsmen were skilled in the process, and the industry continued to grow during the following centuries.

In 1800, Sir Humphry Davy discovered the short-pulse electrical arc and presented his results in 1801. In 1802, Russian scientist Vasily Petrov created the continuous electric arc, and subsequently published "News of Galvanic-Voltaic Experiments" in 1803, in which he described experiments carried out in 1802. Of great importance in this work was the description of a stable arc discharge and the indication of its possible use for many applications, one being melting metals. In 1808, Davy, who was unaware of Petrov's work, rediscovered the continuous electric arc. In 1881–82 inventors Nikolai Benardos (Russian) and Stanisław Olszewski (Polish) created the first electric arc welding method known as carbon arc welding using carbon electrodes. The advances in arc welding continued with the invention of metal electrodes in the late 1800s by a Russian, Nikolai Slavyanov (1888), and an American, C. L. Coffin (1890). Around 1900, A. P. Strohmenger released a coated metal electrode in Britain, which gave a more stable arc. In 1905, Russian scientist Vladimir Mitkevich proposed using a three-phase electric arc for welding. In 1919, alternating current welding was invented by C. J. Holslag but did not become popular for another decade.

Resistance welding was also developed during the final decades of the 19th century, with the first patents going to Elihu Thomson in 1885, who produced further advances over the next 15 years. Thermite welding was invented in 1893, and around that time another process, oxyfuel welding, became well established. Acetylene was discovered in 1836 by Edmund Davy, but its use was not practical in welding until about 1900, when a suitable torch was developed. At first, oxyfuel welding was one of the more popular welding methods due to its portability and relatively low cost. As the 20th century progressed, however, it fell out of favor for industrial applications. It was largely replaced with arc welding, as metal coverings (known as flux) for the electrode that stabilize the arc and shield the base material from impurities continued to be developed.

Bridge of Maurzyce

World War I caused a major surge in the use of welding processes, with the various military powers attempting to determine which of the several new welding processes would be best. The British primarily used arc welding, even constructing a ship, the "Fullagar" with an entirely welded hull. Arc welding was first applied to aircraft during the war as well, as some German airplane fuselages were constructed using the process. Also noteworthy is the first welded road bridge in the world, the Maurzyce Bridge designed by Stefan Bryła of the Lwów University of Technology in 1927, and built across the river Słudwia near Łowicz, Poland in 1928.

Acetylene welding on cylinder water jacket, 1918

During the 1920s, major advances were made in welding technology, including the introduction of automatic welding in 1920, in which electrode wire was fed continuously. Shielding gas became a subject receiving much attention, as scientists attempted to protect welds from the effects of oxygen and nitrogen in the atmosphere. Porosity and brittleness were the primary problems, and the solutions that developed included the use of hydrogen, argon, and helium as welding atmospheres.

During the following decade, further advances allowed for the welding of reactive metals like aluminum and magnesium. This in conjunction with developments in automatic welding, alternating current, and fluxes fed a major expansion of arc welding during the 1930s and then during World War II. In 1930, the first all-welded merchant vessel, M/S Carolinian, was launched.

During the middle of the century, many new welding methods were invented. 1930 saw the release of stud welding, which soon became popular in shipbuilding and construction. Submerged arc welding was invented the same year and continues to be popular today. In 1932 a Russian, Konstantin Khrenov successfully implemented the first underwater electric arc welding. Gas tungsten arc welding, after decades of development, was finally perfected in 1941, and gas metal arc welding followed in 1948, allowing for fast welding of non-ferrous materials but requiring expensive shielding gases. Shielded metal arc welding was developed during the 1950s, using a flux-coated consumable electrode, and it quickly became the most popular metal arc welding process. In 1957, the flux-cored arc welding process debuted, in which the self-shielded wire electrode could be used with automatic equipment, resulting in greatly increased welding speeds, and that same year, plasma arc welding was invented. Electroslag welding was introduced in 1958, and it was followed by its cousin, electrogas welding, in 1961. In 1953 the Soviet scientist N. F. Kazakov proposed the diffusion bonding method.

Other recent developments in welding include the 1958 breakthrough of electron beam welding, making deep and narrow welding possible through the concentrated heat source. Following the invention of the laser in 1960, laser beam welding debuted several decades later, and has proved to be especially useful in high-speed, automated welding. Magnetic pulse welding (MPW) is industrially used since 1967. Friction stir welding was invented in 1991 by Wayne Thomas at The Welding Institute (TWI, UK) and found high-quality applications all over the world. All of these four new processes continue to be quite expensive due the high cost of the necessary equipment, and this has limited their applications.

## Processes

### Arc

Man welding a metal structure in a newly constructed house in Bengaluru, India

These processes use a welding power supply to create and maintain an electric arc between an electrode and the base material to melt metals at the welding point. They can use either direct (DC) or alternating (AC) current, and consumable or non-consumable electrodes. The welding region is sometimes protected by some type of inert or semi-inert gas, known as a shielding gas, and filler material is sometimes used as well.

## Power Supplies

To supply the electrical power necessary for arc welding processes, a variety of different power supplies can be used. The most common welding power supplies are constant current power supplies and constant voltage power supplies. In arc welding, the length of the arc is directly related to the voltage, and the amount of heat input is related to the current. Constant current power supplies are most often used for manual welding processes such as gas tungsten arc welding and shielded metal arc welding, because they maintain a relatively constant current even as the voltage varies. This is important because in manual welding, it can be difficult to hold the electrode perfectly steady, and as a result, the arc length and thus voltage tend to fluctuate. Constant voltage power supplies hold the voltage constant and vary the current, and as a result, are most often used for automated welding processes such as gas metal arc welding, flux cored arc welding, and submerged arc welding. In these processes, arc length is kept constant, since any fluctuation in the distance between the wire and the base material is quickly rectified by a large change in current. For example, if the wire and the base material get too close, the current will rapidly increase, which in turn causes the heat to increase and the tip of the wire to melt, returning it to its original separation distance.

The type of current used plays an important role in arc welding. Consumable electrode processes such as shielded metal arc welding and gas metal arc welding generally use direct current, but the electrode can be charged either positively or negatively. In welding, the positively charged anode will have a greater heat concentration, and as a result, changing the polarity of the electrode affects weld properties. If the electrode is positively charged, the base metal will be hotter, increasing weld penetration and welding speed. Alternatively, a negatively charged electrode results in more shallow welds. Nonconsumable electrode processes, such as gas tungsten arc welding, can use either type of direct current, as well as alternating current. However, with direct current, because the electrode only creates the arc and does not provide filler material, a positively charged electrode causes shallow welds, while a negatively charged electrode makes deeper welds. Alternating current rapidly moves between these two, resulting in medium-penetration welds. One disadvantage of AC, the fact that the arc must be re-ignited after every zero crossing, has been addressed with the invention of special power units that produce a square wave pattern instead of the normal sine wave, making rapid zero crossings possible and minimizing the effects of the problem.

## Processes

One of the most common types of arc welding is shielded metal arc welding (SMAW); it is also known as manual metal arc welding (MMA) or stick welding. Electric current is used to strike an arc between the base material and consumable electrode rod, which is made of filler material (typically steel) and is covered with a flux that protects the weld area from oxidation and contamination by producing carbon dioxide ($CO_2$) gas during the welding process. The electrode core itself acts as filler material, making a separate filler unnecessary.

Shielded metal arc welding

The process is versatile and can be performed with relatively inexpensive equipment, making it well suited to shop jobs and field work. An operator can become reasonably proficient with a modest amount of training and can achieve mastery with experience. Weld times are rather slow, since the consumable electrodes must be frequently replaced and because slag, the residue from the flux, must be chipped away after welding. Furthermore, the process is generally limited to welding ferrous materials, though special electrodes have made possible the welding of cast iron, nickel, aluminum, copper, and other metals.

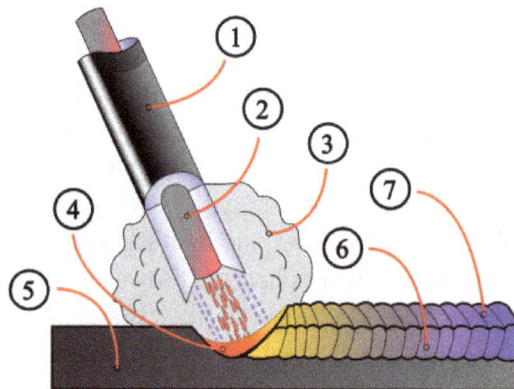

Diagram of arc and weld area, in shielded metal arc welding.

1. Coating Flow

2. Rod

3. Shield Gas

4. Fusion

5. Base metal

6. Weld metal

7. Solidified Slag

Gas metal arc welding (GMAW), also known as metal inert gas or MIG welding, is a semi-automatic or automatic process that uses a continuous wire feed as an electrode and an inert or semi-inert gas mixture to protect the weld from contamination. Since the electrode is continuous, welding speeds are greater for GMAW than for SMAW.

A related process, flux-cored arc welding (FCAW), uses similar equipment but uses wire consisting of a steel electrode surrounding a powder fill material. This cored wire is more expensive than the standard solid wire and can generate fumes and/or slag, but it permits even higher welding speed and greater metal penetration.

Gas tungsten arc welding (GTAW), or tungsten inert gas (TIG) welding, is a manual welding process that uses a nonconsumable tungsten electrode, an inert or semi-inert gas mixture, and a separate filler material. Especially useful for welding thin materials, this method is characterized by a stable arc and high quality welds, but it requires significant operator skill and can only be accomplished at relatively low speeds.

GTAW can be used on nearly all weldable metals, though it is most often applied to stainless steel and light metals. It is often used when quality welds are extremely important, such as in bicycle, aircraft and naval applications. A related process, plasma arc welding, also uses a tungsten electrode but uses plasma gas to make the arc. The arc is more concentrated than the GTAW arc, making transverse control more critical and thus generally restricting the technique to a mechanized process. Because of its stable current, the method can be used on a wider range of material thicknesses than can the GTAW process and it is much faster. It can be applied to all of the same materials as GTAW except magnesium, and automated welding of stainless steel is one important application of the process. A variation of the process is plasma cutting, an efficient steel cutting process.

Submerged arc welding (SAW) is a high-productivity welding method in which the arc is struck beneath a covering layer of flux. This increases arc quality, since contaminants in the atmosphere are blocked by the flux. The slag that forms on the weld generally comes off by itself, and combined with the use of a continuous wire feed, the weld deposition rate is high. Working conditions are much improved over other arc welding processes, since the flux hides the arc and almost no smoke is produced. The process is commonly used in industry, especially for large products and in the manufacture of welded pressure vessels. Other arc welding processes include atomic hydrogen welding, electroslag welding, electrogas welding, and stud arc welding.

## Gas Welding

The most common gas welding process is oxyfuel welding, also known as oxyacetylene welding. It is one of the oldest and most versatile welding processes, but in recent years it has become less popular in industrial applications. It is still widely used for welding pipes and tubes, as well as repair work.

The equipment is relatively inexpensive and simple, generally employing the combustion of acetylene in oxygen to produce a welding flame temperature of about 3100 °C. The flame, since it is less concentrated than an electric arc, causes slower weld cooling, which can lead to greater residual stresses and weld distortion, though it eases the welding of high alloy steels. A similar process, generally called oxyfuel cutting, is used to cut metals.

## Resistance

Resistance welding involves the generation of heat by passing current through the resistance caused by the contact between two or more metal surfaces. Small pools of molten metal are formed at the weld area as high current (1000–100,000 A) is passed through the metal. In general, resistance welding methods are efficient and cause little pollution, but their applications are somewhat limited and the equipment cost can be high.

Spot welder

Spot welding is a popular resistance welding method used to join overlapping metal sheets of up to 3 mm thick. Two electrodes are simultaneously used to clamp the metal sheets together and to pass current through the sheets. The advantages of the method include efficient energy use, limited workpiece deformation, high production rates, easy automation, and no required filler materials. Weld strength is significantly lower than with other welding methods, making the process suitable for only certain applications. It is used extensively in the automotive industry—ordinary cars can have several thousand spot welds made by industrial robots. A specialized process, called shot welding, can be used to spot weld stainless steel.

Like spot welding, seam welding relies on two electrodes to apply pressure and current to join metal sheets. However, instead of pointed electrodes, wheel-shaped electrodes roll along and often feed the workpiece, making it possible to make long continuous welds. In the past, this process was used in the manufacture of beverage cans, but now its uses are more limited. Other resistance welding methods include butt welding, flash welding, projection welding, and upset welding.

## Energy Beam

Energy beam welding methods, namely laser beam welding and electron beam welding, are relatively new processes that have become quite popular in high production applications. The two processes are quite similar, differing most notably in their source of power. Laser beam welding employs a highly focused laser beam, while electron beam welding is done in a vacuum and uses an electron beam. Both have a very high energy density, making deep weld penetration possible and minimizing the size of the weld area. Both processes are extremely fast, and are easily automated, making them highly productive. The primary disadvantages are their very high equipment costs (though these are decreasing) and a susceptibility to thermal cracking. Developments in this area include laser-hybrid welding, which uses principles from both laser beam welding and arc welding for even better weld properties, laser cladding, and x-ray welding.

## Solid-state

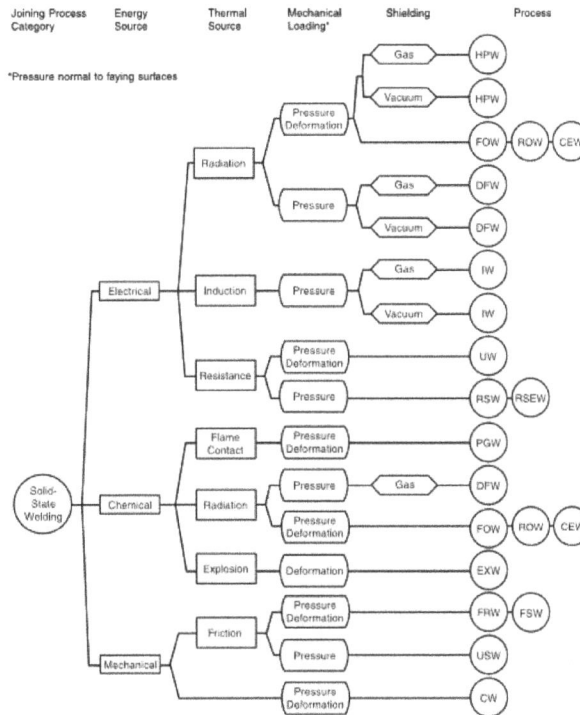

Solid-state welding processes classification chart

Like the first welding process, forge welding, some modern welding methods do not involve the melting of the materials being joined. One of the most popular, ultrasonic welding, is used to connect thin sheets or wires made of metal or thermoplastic by vibrating them at high frequency and under high pressure. The equipment and methods involved are similar to that of resistance welding, but instead of electric current, vibration provides energy input. Welding metals with this process does not involve melting the materials; instead, the weld is formed by introducing mechanical vibrations horizontally under pressure. When welding plastics, the materials should have similar melting temperatures, and the vibrations are introduced vertically. Ultrasonic welding is commonly used for making electrical connections out of aluminum or copper, and it is also a very common polymer welding process.

Another common process, explosion welding, involves the joining of materials by pushing them together under extremely high pressure. The energy from the impact plasticizes the materials, forming a weld, even though only a limited amount of heat is generated. The process is commonly used for welding dissimilar materials, such as the welding of aluminum with steel in ship hulls or compound plates. Other solid-state welding processes include friction welding (including friction stir welding), magnetic pulse welding, co-extrusion welding, cold welding, diffusion bonding, exothermic welding, high frequency welding, hot pressure welding, induction welding, and roll welding.

## Geometry

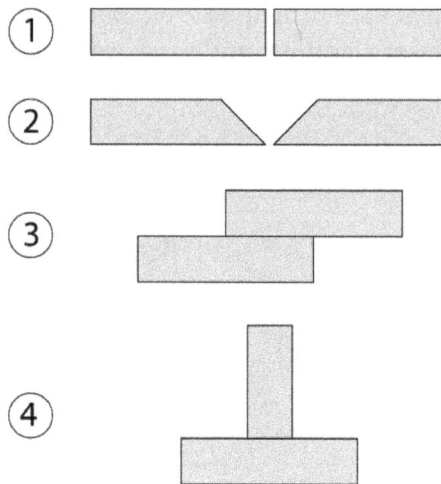

Common welding joint types – (1) Square butt joint, (2) V butt joint, (3) Lap joint, (4) T-joint

Welds can be geometrically prepared in many different ways. The five basic types of weld joints are the butt joint, lap joint, corner joint, edge joint, and T-joint (a variant of this last is the cruciform joint). Other variations exist as well—for example, double-V preparation joints are characterized by the two pieces of material each tapering to a single center point at one-half their height. Single-U and double-U preparation joints are also fairly common—instead of having straight edges like the single-V and double-V preparation joints, they are curved, forming the shape of a U. Lap joints are also commonly more than two pieces thick—depending on the process used and the thickness of the material, many pieces can be welded together in a lap joint geometry.

Many welding processes require the use of a particular joint design; for example, resistance spot welding, laser beam welding, and electron beam welding are most frequently performed on lap joints. Other welding methods, like shielded metal arc welding, are extremely versatile and can weld virtually any type of joint. Some processes can also be used to make multipass welds, in which one weld is allowed to cool, and then another weld is performed on top of it. This allows for the welding of thick sections arranged in a single-V preparation joint, for example.

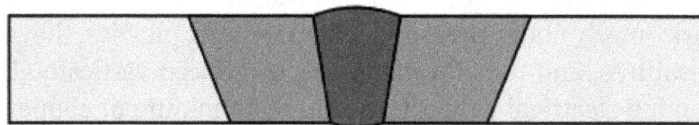

The cross-section of a welded butt joint, with the darkest gray representing the weld or fusion zone, the medium gray the heat-affected zone, and the lightest gray the base material.

After welding, a number of distinct regions can be identified in the weld area. The weld itself is called the fusion zone—more specifically, it is where the filler metal was laid during the welding process. The properties of the fusion zone depend primarily on the filler metal used, and its compatibility with the base materials. It is surrounded by the heat-affected zone, the area that had its microstructure and properties altered by the weld. These properties depend on the base material's behavior when subjected to heat. The metal in this area is often weaker than both the base material and the fusion zone, and is also where residual stresses are found.

## Quality

The blue area results from oxidation at a corresponding temperature of 600 °F (316 °C). This is an accurate way to identify temperature, but does not represent the HAZ width. The HAZ is the narrow area that immediately surrounds the welded base metal.

Many distinct factors influence the strength of welds and the material around them, including the welding method, the amount and concentration of energy input, the weldability of the base material, filler material, and flux material, the design of the joint, and the interactions between all these factors. To test the quality of a weld, either destructive or nondestructive testing methods are commonly used to verify that welds are free of defects, have acceptable levels of residual stresses and distortion, and have acceptable heat-affected zone (HAZ) properties. Types of welding defects include cracks, distortion, gas inclusions (porosity), non-metallic inclusions, lack of fusion, incomplete penetration, lamellar tearing, and undercutting.

The metalworking industry has instituted specifications and codes to guide welders, weld inspectors, engineers, managers, and property owners in proper welding technique, design of welds, how to judge the quality of Welding Procedure Specification, how to judge the skill of the person performing the weld, and how to ensure the quality of a welding job. Methods such as visual inspection, radiography, ultrasonic testing, phased-array ultrasonics, dye penetrant inspection, magnetic particle inspection, or industrial computed tomography can help with detection and analysis of certain defects.

## Heat-affected Zone

The effects of welding on the material surrounding the weld can be detrimental—depending on the materials used and the heat input of the welding process used, the HAZ can be of varying size and strength. The thermal diffusivity of the base material plays a large role—if the diffusivity is high, the material cooling rate is high and the HAZ is relatively small. Conversely, a low diffusivity leads to slower cooling and a larger HAZ. The amount of heat injected by the welding process plays an important role as well, as processes like oxyacetylene welding have an unconcentrated heat input and increase the size of the HAZ. Processes like laser beam welding give a highly concentrated, limited amount of heat, resulting in a small HAZ. Arc welding falls between these two extremes, with the individual processes varying somewhat in heat input. To calculate the heat input for arc welding procedures, the following formula can be used:

$$Q = \left( \frac{V \times I \times 60}{S \times 1000} \right) \times \textit{Efficiency}$$

where $Q$ = heat input (kJ/mm), $V$ = voltage (V), $I$ = current (A), and $S$ = welding speed (mm/min). The efficiency is dependent on the welding process used, with shielded metal arc welding having a value of 0.75, gas metal arc welding and submerged arc welding, 0.9, and gas tungsten arc welding, 0.8.

## Lifetime Extension with Aftertreatment Methods

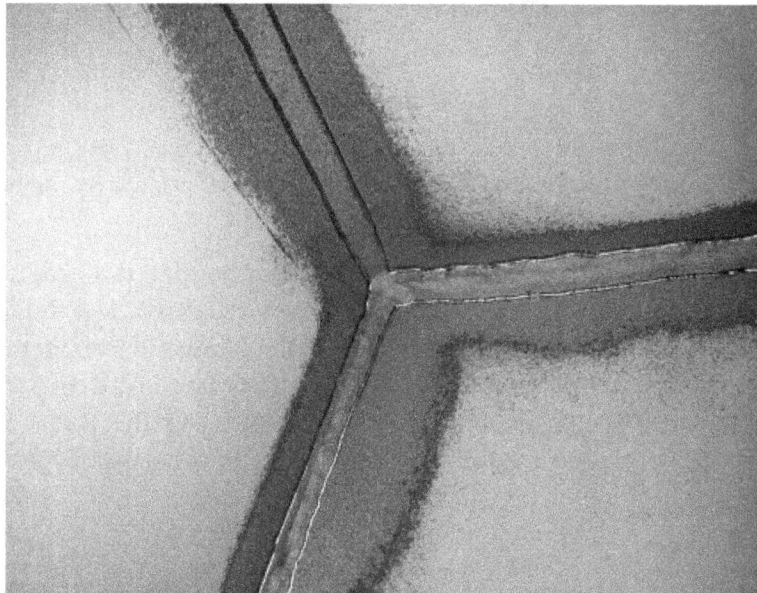

Example: High Frequency Impact Treatment for lifetime extension

The durability and life of dynamically loaded, welded steel structures is determined in many cases by the welds, particular the weld transitions. Through selective treatment of the transitions by grinding (abrasive cutting), shot peening, High Frequency Impact Treatment, etc. the durability of many designs increase significantly.

## Metallurgy

Most solids used are engineering materials consisting of crystalline solids in which the atoms or ions are arranged in a repetitive geometric pattern which is known as a lattice structure. The only exception is material that is made from glass which is a combination of a supercooled liquid and polymers which are aggregates of large organic molecules.

Crystalline solids cohesion is obtained by a metallic or chemical bond which is formed between the constituent atoms. Chemical bonds can be grouped into two types consisting of ionic and co-valent. To form an ionic bond, either a valence or bonding electron separates from one atom and becomes attached to another atom to form oppositely charged ions. The bonding in the static position is when the ions occupy an equilibrium position where the resulting force between them is zero. When the ions are exerted in tension force, the inter-ionic spacing increases creating an electrostatic attractive force, while a repulsing force under compressive force between the atomic nuclei is dominant.

Covalent bonding takes place when one of the constituent atoms loses one or more electrons, with the other atom gaining the electrons, resulting in an electron cloud that is shared by the molecule as a whole. In both ionic and covalent bonding the location of the ions and electrons are con-strained relative to each other, thereby resulting in the bond being characteristically brittle.

Metallic bonding can be classified as a type of covalent bonding for which the constituent atoms of the same type and do not combine with one another to form a chemical bond. Atoms will lose an electron(s) forming an array of positive ions. These electrons are shared by the lattice which makes the electron cluster mobile, as the electrons are free to move as well as the ions. For this, it gives metals their relatively high thermal and electrical conductivity as well as being characteristically ductile.

Three of the most commonly used crystal lattice structures in metals are the body-centred cubic, face-centred cubic and close-packed hexagonal. Ferritic steel has a body-centred cubic structure and austenitic steel, non-ferrous metals like aluminum, copper and nickel have the face-centred cubic structure.

Ductility is an important factor in ensuring the integrity of structures by enabling them to sustain local stress concentrations without fracture. In addition, structures are required to be of an accept-able strength, which is related to a material's yield strength. In general, as the yield strength of a material increases, there is a corresponding reduction in fracture toughness.

A reduction in fracture toughness may also be attributed to the embrittlement effect of impurities, or for body-centred cubic metals, from a reduction in temperature. Metals and in particular steels have a transitional temperature range where above this range the metal has acceptable notch-duc-tility while below this range the material becomes brittle. Within the range, the materials behavior is unpredictable. The reduction in fracture toughness is accompanied by a change in the fracture appearance. When above the transition, the fracture is primarily due to micro-void coalescence, which results in the fracture appearing fibrous. When the temperatures falls the fracture will show signs of cleavage facets. These two appearances are visible by the naked eye. Brittle fracture in steel plates may appear as chevron markings under the microscope. These arrow-like ridges on the crack surface point towards the origin of the fracture.

Fracture toughness is measured using a notched and pre-cracked rectangular specimen, of which the dimensions are specified in standards, for example ASTM E23. There are other means of esti-mating or measuring fracture toughness by the following: The Charpy impact test per ASTM A370; The crack-tip opening displacement (CTOD) test per BS 7448-1; The J integral test per ASTM E1820; The Pellini drop-weight test per ASTM E208.

## Unusual Conditions

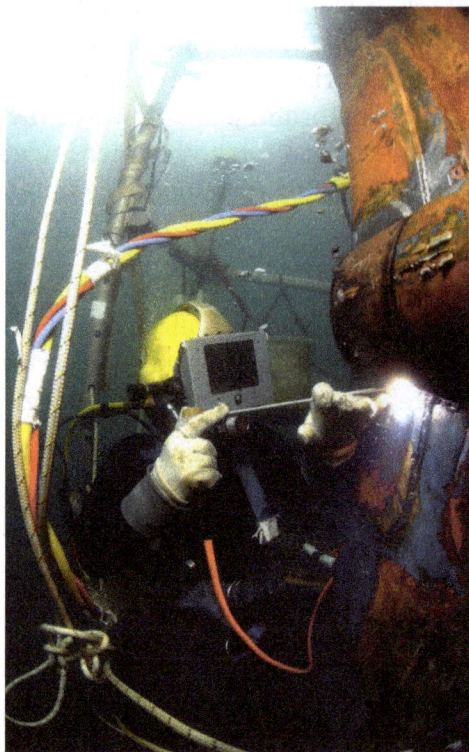

Underwater welding

While many welding applications are done in controlled environments such as factories and repair shops, some welding processes are commonly used in a wide variety of conditions, such as open air, underwater, and vacuums (such as space). In open-air applications, such as construction and outdoors repair, shielded metal arc welding is the most common process. Processes that employ inert gases to protect the weld cannot be readily used in such situations, because unpredictable atmospheric movements can result in a faulty weld. Shielded metal arc welding is also often used in underwater welding in the construction and repair of ships, offshore platforms, and pipelines, but others, such as flux cored arc welding and gas tungsten arc welding, are also common. Welding in space is also possible—it was first attempted in 1969 by Russian cosmonauts, when they performed experiments to test shielded metal arc welding, plasma arc welding, and electron beam welding in a depressurized environment. Further testing of these methods was done in the following decades, and today researchers continue to develop methods for using other welding processes in space, such as laser beam welding, resistance welding, and friction welding. Advances in these areas may be useful for future endeavours similar to the construction of the International Space Station, which could rely on welding for joining in space the parts that were manufactured on Earth.

## Safety Issues

Welding can be dangerous and unhealthy if the proper precautions are not taken. However, using new technology and proper protection greatly reduces risks of injury and death associated with welding. Since many common welding procedures involve an open electric arc or flame, the risk of burns and fire is significant; this is why it is classified as a hot work process. To prevent injury, welders wear personal protective equipment in the form of heavy leather gloves and protective long-sleeve jackets to avoid exposure to extreme heat and flames. Additionally, the brightness of the weld area leads to a condition called arc eye or flash burns in which ultraviolet light causes inflammation of the cornea and can burn the retinas of the eyes. Goggles and welding helmets with dark UV-filtering face plates are worn to prevent this exposure. Since the 2000s, some helmets have included a face plate which instantly darkens upon exposure to the intense UV light. To protect bystanders, the welding area is often surrounded with translucent welding curtains. These curtains, made of a polyvinyl chloride plastic film, shield people outside the welding area from the UV light of the electric arc, but can not replace the filter glass used in helmets.

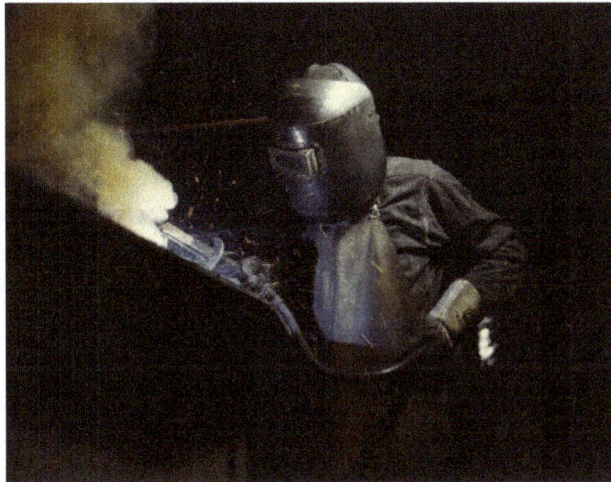

Arc welding with a welding helmet, gloves, and other protective clothing

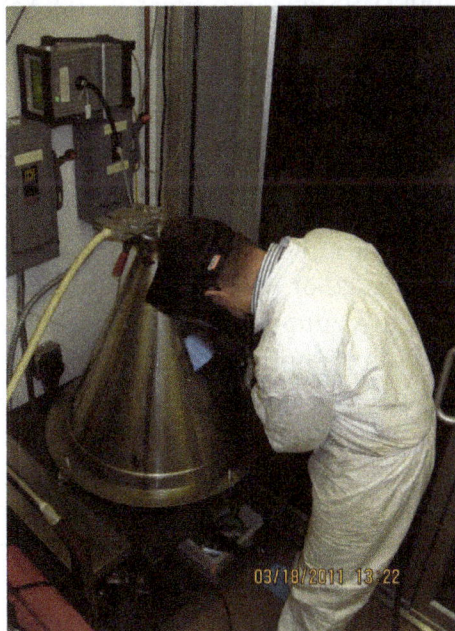

A chamber designed to contain welding fumes for analysis

Welders are often exposed to dangerous gases and particulate matter. Processes like flux-cored arc welding and shielded metal arc welding produce smoke containing particles of various types of oxides. The size of the particles in question tends to influence the toxicity of the fumes, with smaller particles presenting a greater danger. This is because smaller particles have the ability to cross the blood brain barrier. Fumes and gases, such as carbon dioxide, ozone, and fumes containing heavy metals, can be dangerous to welders lacking proper ventilation and training. Exposure to manganese welding fumes, for example, even at low levels ($<0.2$ mg/m$^3$), may lead to neurological problems or to damage to the lungs, liver, kidneys, or central nervous system. Nano particles can become trapped in the alveolar macrophages of the lungs and induce pulmonary fibrosis. The use of compressed gases and flames in many welding processes poses an explosion and fire risk. Some common precautions include limiting the amount of oxygen in the air, and keeping combustible materials away from the workplace.

## Costs and Trends

As an industrial process, the cost of welding plays a crucial role in manufacturing decisions. Many different variables affect the total cost, including equipment cost, labor cost, material cost, and energy cost. Depending on the process, equipment cost can vary, from inexpensive for methods like shielded metal arc welding and oxyfuel welding, to extremely expensive for methods like laser beam welding and electron beam welding. Because of their high cost, they are only used in high production operations. Similarly, because automation and robots increase equipment costs, they are only implemented when high production is necessary. Labor cost depends on the deposition rate (the rate of welding), the hourly wage, and the total operation time, including time spent fitting, welding, and handling the part. The cost of materials includes the cost of the base and filler material, and the cost of shielding gases. Finally, energy cost depends on arc time and welding power demand.

For manual welding methods, labor costs generally make up the vast majority of the total cost. As a result, many cost-saving measures are focused on minimizing operation time. To do this, welding procedures with high deposition rates can be selected, and weld parameters can be fine-tuned to increase welding speed. Mechanization and automation are often implemented to reduce labor costs, but this frequently increases the cost of equipment and creates additional setup time. Material costs tend to increase when special properties are necessary, and energy costs normally do not amount to more than several percent of the total welding cost.

In recent years, in order to minimize labor costs in high production manufacturing, industrial welding has become increasingly more automated, most notably with the use of robots in resistance spot welding (especially in the automotive industry) and in arc welding. In robot welding, mechanized devices both hold the material and perform the weld and at first, spot welding was its most common application, but robotic arc welding increases in popularity as technology advances. Other key areas of research and development include the welding of dissimilar materials (such as steel and aluminum, for example) and new welding processes, such as friction stir, magnetic pulse, conductive heat seam, and laser-hybrid welding. Furthermore, progress is desired in making more specialized methods like laser beam welding practical for more applications, such as in the aerospace and automotive industries. Researchers also hope to better understand the often unpredictable properties of welds, especially microstructure, residual stresses, and a weld's tendency to crack or deform.

The trend of accelerating the speed at which welds are performed in the steel erection industry comes at a risk to the integrity of the connection. Without proper fusion to the base materials provided by sufficient arc time on the weld, a project inspector cannot ensure the effective diameter of the puddle weld therefore he or she cannot guarantee the published load capacities unless they witness the actual installation. This method of puddle welding is common in the United States and Canada for attaching steel sheets to bar joist and structural steel members. Regional agencies are responsible for ensuring the proper installation of puddle welding on steel construction sites. Currently there is no standard or weld procedure which can ensure the published holding capacity of any unwitnessed connection, but this is under review by the American Welding Society.

## Glass and Plastic Welding

The welding together of two tubes made from lead glass

Glasses and certain types of plastics are commonly welded materials. Unlike metals, which have a specific melting point, glasses and plastics have a melting range, called the glass transition. When heating the solid material into this range, it will generally become softer and more pliable. When it crosses through the glass transition, it will become a very thick, sluggish, viscous liquid. Typically, this viscous liquid will have very little surface tension, becoming a sticky, honey-like consistency, so welding can usually take place by simply pressing two melted surfaces together. The two liquids will generally mix and join at first contact. Upon cooling through the glass transition, the welded piece will solidify as one solid piece of amorphous material.

A bowl made from cast-glass. The two halves are joined together by the weld seam, running down the middle.

## Glass Welding

Glass welding is a common practice during glassblowing. It is used very often in the construction of lighting, neon signs, flashtubes, scientific equipment, and the manufacture of dishes and other glassware. It is also used during glass casting for joining the halves of glass molds, making items such as bottles and jars. Welding glass is accomplished by heating the glass through the glass transition, turning it into a thick, formable, liquid mass. Heating is usually done with a gas or oxy-gas torch, or a furnace, because the temperatures for melting glass are often quite high. This temperature may vary, depending on the type of glass. For example, lead glass becomes a weldable liquid at around 1,600 °F (870 °C), and can be welded with a simple propane torch. On the other hand, quartz glass (fused silica) must be heated to over 3,000 °F (1,650 °C), but quickly loses its viscosity and formability if overheated, so an oxyhydrogen torch must be used. Sometimes a tube may be attached to the glass, allowing it to be blown into various shapes, such as bulbs, bottles, or tubes. When two pieces of liquid glass are pressed together, they will usually weld very readily. Welding a handle onto a pitcher can usually be done with relative ease. However, when welding a tube to another tube, a combination of blowing and suction, and pressing and pulling is used to ensure a good seal, to shape the glass, and to keep the surface tension from closing the tube in on itself. Sometimes a filler rod may be used, but usually not.

Because glass is very brittle in its solid state, it is often prone to cracking upon heating and cooling, especially if the heating and cooling are uneven. This is because the brittleness of glass does not allow for uneven thermal expansion. Glass that has been welded will usually need to be cooled very slowly and evenly through the glass transition, in a process called annealing, to relieve any internal stresses created by a temperature gradient.

There are many types of glass, and it is most common to weld using the same types. Different glasses often have different rates of thermal expansion, which can cause them to crack upon cooling when they contract differently. For instance, quartz has very low thermal expansion, while soda-lime glass has very high thermal expansion. When welding different glasses to each other, it is usually important to closely match their coefficients of thermal expansion, to ensure that cracking does not occur. Also, some glasses will simply not mix with others, so welding between certain types may not be possible.

Glass can also be welded to metals and ceramics, although with metals the process is usually more adhesion to the surface of the metal rather than a commingling of the two materials. However, certain glasses will typically bond only to certain metals. For example, lead glass bonds readily to copper or molybdenum, but not to aluminum. Tungsten electrodes are often used in lighting but will not bond to quartz glass, so the tungsten is often wetted with molten borosilicate glass, which bonds to both tungsten and quartz. However, care must be taken to ensure that all materials have similar coefficients of thermal expansion to prevent cracking both when the object cools and when it is heated again. Special alloys are often used for this purpose, ensuring that the coefficients of expansion match, and sometimes thin, metallic coatings may be applied to a metal to create a good bond with the glass.

## Plastic Welding

Plastics are generally divided into two categories, which are "thermosets" and "thermoplastics." A thermoset is a plastic in which a chemical reaction sets the molecular bonds after first forming the plastic, and then the bonds cannot be broken again without degrading the plastic. Thermosets cannot be melted, therefore, once a thermoset has set it is impossible to weld it. Examples of thermosets include epoxies, silicone, vulcanized rubber, polyester, and polyurethane.

Thermoplastics, by contrast, form long molecular chains, which are often coiled or intertwined, forming an amorphous structure without any long-range, crystalline order. Some thermoplastics may be fully amorphous, while others have a partially crystalline/partially amorphous structure. Both amorphous and semicrystalline thermoplastics have a glass transition, above which welding can occur, but semicrystallines also have a specific melting point which is above the glass transition. Above this melting point, the viscous liquid will become a free-flowing liquid. Examples of thermoplastics include polyethylene, polypropylene, polystyrene, polyvinylchloride (PVC), and fluoroplastics like Teflon and Spectralon.

Welding thermoplastic is very similar to welding glass. The plastic first must be cleaned and then heated through the glass transition, turning the weld-interface into a thick, viscous liquid. Two heated interfaces can then be pressed together, allowing the molecules to mix through intermolecular diffusion, joining them as one. Then the plastic is cooled through the glass transition, allowing the weld to solidify. A filler rod may often be used for certain types of joints. The main differences between welding glass and plastic are the types of heating methods, the much lower melting temperatures, and the fact that plastics will burn if overheated. Many different methods have been devised for heating plastic to a weldable temperature without burning it. Ovens or electric heating tools can be used to melt the plastic. Ultrasonic, laser, or friction heating are other methods. Resistive metals may be implanted in the plastic, which respond to induction heating. Some plastics will begin to burn at temperatures lower than their glass transition, so welding can be performed by blowing a heated, inert gas onto the plastic, melting it while, at the same time, shielding it from oxygen.

Many thermoplastics can also be welded using chemical solvents. When placed in contact with the plastic, the solvent will begin to soften it, bringing the surface into a thick, liquid solution. When two melted surfaces are pressed together, the molecules in the solution mix, joining them as one. Because the solvent can permeate the plastic, the solvent evaporates out through the surface of the plastic, causing the weld to drop out of solution and solidify. A common use for solvent welding is for joining PVC or ABS (acrylonitrile butadiene styrene) pipes during plumbing, or for welding styrene and polystyrene plastics in the construction of models. Solvent welding is especially effective on plastics like PVC which burn at or below their glass transition, but may be ineffective on plastics like Teflon or polyethylene that are resistant to chemical decomposition.

## Annealing (Metallurgy)

Annealing, in metallurgy and materials science, is a heat treatment that alters the physical and sometimes chemical properties of a material to increase its ductility and reduce its hardness, making it more workable. It involves heating a material to above its recrystallization temperature, maintaining a suitable temperature, and then cooling.

In annealing, atoms migrate in the crystal lattice and the number of dislocations decreases, leading to the change in ductility and hardness.

In the cases of copper, steel, silver, and brass, this process is performed by heating the material (generally until glowing) for a while and then slowly letting it cool to room temperature in still air. Copper, silver and brass can be cooled slowly in air, or quickly by quenching in water, unlike ferrous metals, such as steel, which must be cooled slowly to anneal. In this fashion, the metal is softened and prepared for further work—such as shaping, stamping, or forming.

## Thermodynamics

Annealing occurs by the diffusion of atoms within a solid material, so that the material progresses towards its equilibrium state. Heat increases the rate of diffusion by providing the energy needed to break bonds. The movement of atoms has the effect of redistributing and eradicating the dislocations in metals and (to a lesser extent) in ceramics. This alteration to existing dislocations allows a metal object to deform more easily, increasing its ductility.

The amount of process-initiating Gibbs free energy in a deformed metal is also reduced by the annealing process. In practice and industry, this reduction of Gibbs free energy is termed *stress relief*.

The relief of internal stresses is a thermodynamically spontaneous process; however, at room temperatures, it is a very slow process. The high temperatures at which annealing occurs serve to accelerate this process.

The reaction that facilitates returning the cold-worked metal to its stress-free state has many reaction pathways, mostly involving the elimination of lattice vacancy gradients within the body of the metal. The creation of lattice vacancies is governed by the Arrhenius equation, and the migration/diffusion of lattice vacancies are governed by Fick's laws of diffusion.

In steel, there is a decarburation mechanism that can be described as three distinct events: the reaction at the steel surface, the interstitial diffusion of carbon atoms and the dissolution of carbides within the steel.

## Stages

The three stages of the annealing process that proceed as the temperature of the material is increased are: recovery, recrystallization, and grain growth. The first stage is recovery, and it results in softening of the metal through removal of primarily linear defects called *dislocations* and the internal stresses they cause. Recovery occurs at the lower temperature stage of all annealing processes and before the appearance of new strain-free grains. The grain size and shape do not change. The second stage is recrystallization, where new strain-free grains nucleate and grow to replace those deformed by internal stresses. If annealing is allowed to continue once recrystallization has completed, then grain growth (the third stage) occurs. In grain growth, the microstructure starts to coarsen and may cause the metal to lose a substantial part of its original strength. This can however be regained with hardening.

## Controlled Atmospheres

The high temperature of annealing may result in oxidation of the metal's surface, resulting in scale. If scale must be avoided, annealing is carried out in a special atmosphere, such as with endothermic gas (a mixture of carbon monoxide, hydrogen gas, and nitrogen gas). Annealing is also done in forming gas, a mixture of hydrogen and nitrogen.

The magnetic properties of mu-metal (Espey cores) are introduced by annealing the alloy in a hydrogen atmosphere.

## Setup and Equipment

Typically, large ovens are used for the annealing process. The inside of the oven is large enough to place the workpiece in a position to receive maximum exposure to the circulating heated air. For high volume process annealing, gas fired conveyor furnaces are often used. For large workpieces or high quantity parts, car-bottom furnaces are used so workers can easily move the parts in and out. Once the annealing process is successfully completed, workpieces are sometimes left in the oven so the parts cool in a controllable way. While some workpieces are left in the oven to cool in a controlled fashion, other materials and alloys are removed from the oven. Once removed from the oven, the workpieces are often quickly cooled off in a process known as quench hardening. Typical methods of quench hardening materials involve media such as air, water, oil, or salt. Salt is used as a medium for quenching usually in the form of brine (salt water). Brine provides faster cooling rates than water. This is because when an object is quenched in water air bubbles form on the surface of the object reducing the surface area the water is in contact with. The salt in the brine reduces the formation of air bubbles on the object's surface, meaning there is a larger surface area of the object in contact with the water, providing faster cooling rates. Quench hardening is generally applicable to some ferrous alloys, but not copper alloys.

## Diffusion Annealing of Semiconductors

In the semiconductor industry, silicon wafers are annealed, so that dopant atoms, usually boron, phosphorus or arsenic, can diffuse into substitutional positions in the crystal lattice, resulting in drastic changes in the electrical properties of the semiconducting material.

## Specialized Cycles

## Normalization

*Normalization* is an annealing process applied to ferrous alloys to give the material a uniform fine-grained structure and make it less brittle. It is used on steels of less than 0.4% carbon to transform austenite into ferrite, pearlite and sorbite. It involves heating the steel to 20-50 Kelvin above its upper critical point. It is soaked for a short period at that temperature and then allowed to cool in air. Smaller grains form that produce a tougher, more ductile material. It eliminates columnar grains and dendritic segregation that sometimes occurs during casting. Normalizing improves machinability of a component and provides dimensional stability if subjected to further heat treatment processes.

## Process Annealing

Process annealing, also called *intermediate annealing, subcritical annealing,* or *in-process annealing,* is a heat treatment cycle that restores some of the ductility to a product being cold-worked so it can be cold-worked further without breaking.

The temperature range for process annealing ranges from 260 °C (500 °F) to 760 °C (1400 °F), depending on the alloy in question. This process is mainly suited for low-carbon steel. The material is heated up to a temperature just below the lower critical temperature of steel. Cold-worked steel normally tends to possess increased hardness and decreased ductility, making it difficult to work. Process annealing tends to improve these characteristics. This is mainly carried out on cold-rolled steel like wire-drawn steel, etc.

## Full Anneal

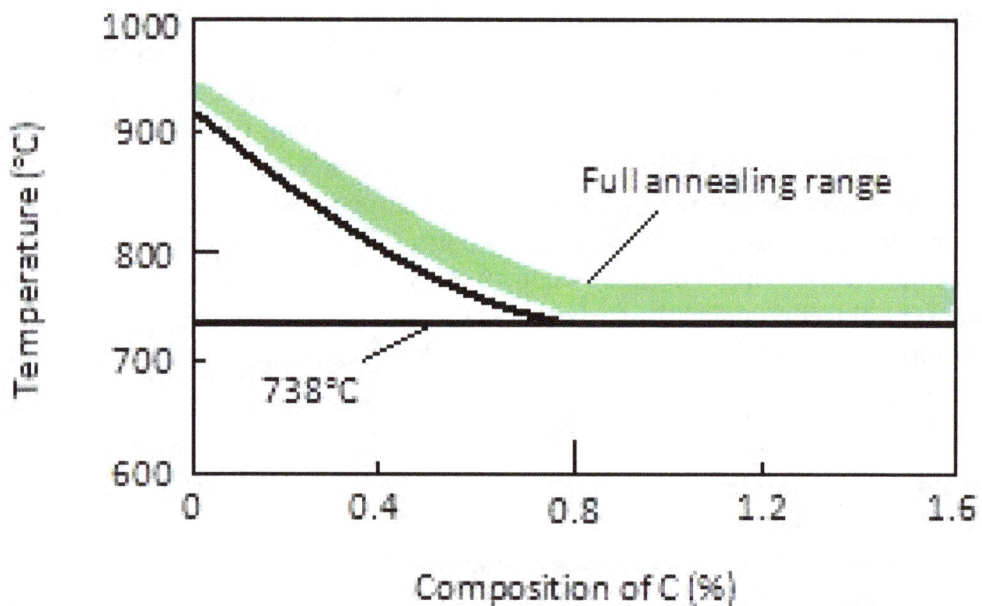

Full annealing temperature ranges

A full anneal typically results in the second most ductile state a metal can assume for metal alloy. Its purpose is to originate a uniform and stable microstructure that most closely resembles the metal's phase diagram equilibrium microstructure, thus letting the metal attain relatively low levels of hardness, yield strength and ultimate strength with high plasticity and toughness. To perform a full anneal on a steel for example, steel is heated to slightly above the austenitic temperature and held for sufficient time to allow the material to fully form austenite or austenite-cementite grain structure. The material is then allowed to cool very slowly so that the equilibrium microstructure is obtained. In most cases this means the material is allowed to furnace cool (the furnace is turned off and the steel is let cool down inside) but in some cases it's air cooled. The cooling rate of the steel has to be sufficiently slow so as to not let the austenite transform into bainite or martensite, but rather have it completely transform to pearlite and ferrite or cementite. This means that steels that are very hardenable (i.e. tend to form martensite under moderately low cooling rates) have to be furnace cooled. The details of the process depend on the type of metal and the precise alloy involved. In any case the result is a more ductile material but a lower yield strength and a lower tensile strength. This process is also called LP annealing for *lamellar pearlite* in the steel industry as opposed to a *process anneal*, which does not specify a microstructure and only has the goal of softening the material. Often the material to be machined is annealed, and then subject to further heat treatment to achieve the final desired properties.

## Short Cycle Anneal

Short cycle annealing is used for turning normal ferrite into malleable ferrite. It consists of heating, cooling and then heating again from 4 to 8 hours.

## Resistive Heating

Resistive heating can be used to efficiently anneal copper wire; the heating system employs a controlled electrical short circuit. It can be advantageous because it does not require a temperature-regulated furnace like other methods of annealing.

The process consists of two conductive pulleys (*step pulleys*), which the wire passes across after it is drawn. The two pulleys have an electrical potential across them, which causes the wire to form a short circuit. The Joule effect causes the temperature of the wire to rise to approximately 400 °C. This temperature is affected by the rotational speed of the pulleys, the ambient temperature, and the voltage applied. Where t is the temperature of the wire, K is a constant, V is the voltage applied, r is the number of rotations of the pulleys per minute, and $t_a$ is the ambient temperature:

$$t = \frac{1}{r}KV^2 + t_a$$

The constant K depends on the diameter of the pulleys and the resistivity of the copper.

Purely in terms of the temperature of the copper wire, an increase in the speed of the wire through the pulley system has the same effect as an increase in resistance. Therefore, the speed of the wire can be varied quadratically as the voltage is applied.

# Calcination

Authorities differ on the meaning of calcination (also referred to as calcining). The IUPAC defines it as 'Heating to high temperatures in air or oxygen'. However calcination is also used to mean a thermal treatment process in the absence or limited supply of air or oxygen applied to ores and other solid materials to bring about a thermal decomposition A calciner is a steel cylinder that rotates inside a heated furnace and performs indirect high-temperature processing (550-1150 °C, or 1000-2100 °F) within a controlled atmosphere.

## Industrial Processes

The process of calcination derives its name from the Latin *calcinare* (to burn lime) due to its most common application, the decomposition of calcium carbonate (limestone) to calcium oxide (lime) and carbon dioxide, in order to create cement. The product of calcination is usually referred to in general as "calcine", regardless of the actual minerals undergoing thermal treatment. Calcination is carried out in furnaces or reactors (sometimes referred to as kilns or calciners) of various designs including shaft furnaces, rotary kilns, multiple hearth furnaces, and fluidized bed reactors.

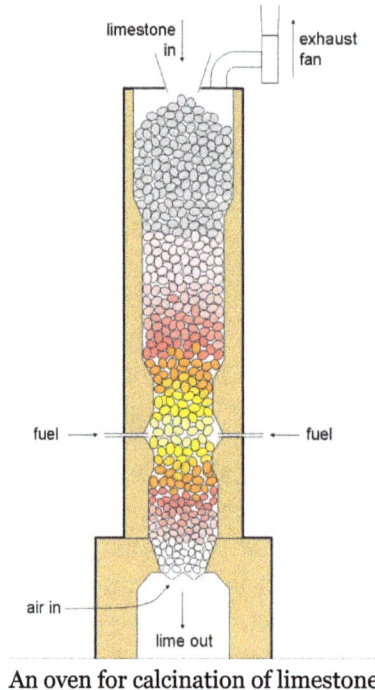

An oven for calcination of limestone

Examples of calcination processes include the following:

- decomposition of carbonate minerals, as in the calcination of limestone to drive off carbon dioxide;

- decomposition of hydrated minerals, as in the calcination of bauxite and gypsum, to remove crystalline water as water vapor;

- decomposition of volatile matter contained in raw petroleum coke;

- heat treatment to effect phase transformations, as in conversion of anatase to rutile or devitrification of glass materials

- removal of ammonium ions in the synthesis of zeolites.

## Calcination Reactions

Calcination reactions usually take place at or above the thermal decomposition temperature (for decomposition and volatilization reactions) or the transition temperature (for phase transitions). This temperature is usually defined as the temperature at which the standard Gibbs free energy for a particular calcination reaction is equal to zero. For example, in limestone calcination, a decomposition process, the chemical reaction is

$$CaCO_3 \rightarrow CaO + CO_2(g)$$

The standard Gibbs free energy of reaction is approximated as $\Delta G°_r = 177,100 - 158\ T$ (J/mol). The standard free energy of reaction is 0 in this case when the temperature, $T$, is equal to 1121 K, or 848 °C.

## Oxidation

In some cases, calcination of a metal results in oxidation of the metal. Jean Rey noted that lead and tin when calcinated gained weight, presumably as they were being oxidized.

## Alchemy

In alchemy, calcination was believed to be one of the 12 vital processes required for the transformation of a substance.

Alchemists distinguished two kinds of calcination, *actual* and *potential*. Actual calcination is that brought about by actual fire, from wood, coals, or other fuel, raised to a certain temperature. Potential calcination is that brought about by *potential* fire, such as corrosive chemicals; for example, gold was calcined in a reverberatory furnace with mercury and sal ammoniac; silver with common salt and alkali salt; copper with salt and sulfur; iron with sal ammoniac and vinegar; tin with antimony; lead with sulfur; and mercury with aqua fortis.

There was also *philosophical calcination*, which was said to occur when horns, hooves, etc., were hung over boiling water, or other liquor, until they had lost their mucilage, and were easily reducible into powder.

# Recovery (Metallurgy)

**Recovery** is a process by which deformed grains can reduce their stored energy by the removal or rearrangement of defects in their crystal structure. These defects, primarily dislocations, are introduced by plastic deformation of the material and act to increase the yield strength of a material. Since recovery reduces the dislocation density the process is normally accompanied by a reduction in a materials strength and a simultaneous increase in the ductility. As a result, recovery may be considered beneficial or detrimental depending on the circumstances. Recovery is related to the similar process of recrystallization and grain growth. Recovery competes with recrystallization, as both are driven by the stored energy, but is also thought to be a necessary prerequisite for the nucleation of recrystallized grains. It is so called because there is a recovery of the electrical conductivity due to a reduction in dislocations. This creates defect-free channels, giving electrons an increased mean-free path.

## Definition

The physical processes that fall under the designations of recovery, recrystallisation and grain growth are often difficult to distinguish in a precise manner. Doherty *et al.* (1998) stated:

"The authors have agreed that ... recovery can be defined as all annealing processes occurring in deformed materials that occur without the migration of a high-angle grain boundary"

Thus the process can be differentiated from recrystallisation and grain growth as both feature extensive movement of high-angle grain boundaries.

If recovery occurs during deformation (a situation that is common in high-temperature processing) then it is referred to as 'dynamic' while recovery that occurs after processing is termed 'static'. The principal difference is that during dynamic recovery, stored energy continues to be introduced even as it is decreased by the recovery process - resulting in a form of dynamic equilibrium.

## Process

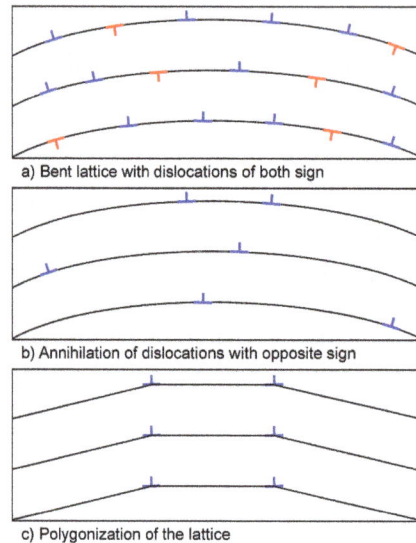

a) Bent lattice with dislocations of both sign

b) Annihilation of dislocations with opposite sign

c) Polygonization of the lattice

The annihilation and reorganisation of an array of edge dislocations in a crystal lattice

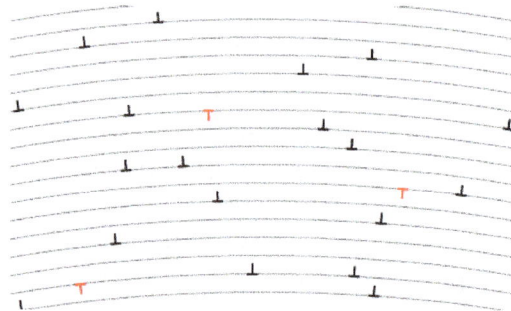

Animation of the annihilation and reorganisation of edge dislocations in a crystal lattice

## Deformed Structure

A heavily deformed metal contains a huge number of dislocations predominantly caught up in 'tangles' or 'forests'. Dislocation motion is relatively difficult in a metal with a low stacking fault energy and so the dislocation distribution after deformation is largely random. In contrast, metals with moderate to high stacking fault energy, e.g. aluminum, tend to form a cellular structure where the cell walls consist of rough tangles of dislocations. The interiors of the cells have a correspondingly reduced dislocation density.

## Annihilation

Each dislocation is associated with a strain field which contributes some small but finite amount to the materials stored energy. When the temperature is increased - typically below one-third of the absolute melting point - dislocations become mobile and are able to glide, cross-slip and climb. If two dislocations of opposite sign meet then they effectively cancel out and their contribution to the stored energy is removed. When annihilation is complete then only excess dislocation of one kind will remain.

## Rearrangement

After annihilation any remaining dislocations can align themselves into ordered arrays where their individual contribution to the stored energy is reduced by the overlapping of their strain fields. The simplest case is that of an array of edge dislocations of identical Burger's vector. This idealised case can be produced by bending a single crystal that will deform on a single slip system (the original experiment performed by Cahn in 1949). The edge dislocations will rearrange themselves into tilt boundaries, a simple example of a low-angle grain boundary. Grain boundary theory predicts that an increase in boundary misorientation will increase the energy of the boundary but decrease the energy per dislocation. Thus, there is a driving force to produce fewer, more highly misoriented boundaries. The situation in highly deformed, polycrystalline materials is naturally more complex. Many dislocations of different Burger's vector can interact to form complex 2-D networks.

## Development of Substructure

As mentioned above, the deformed structure is often a 3-D cellular structure with walls consisting of dislocation tangles. As recovery proceeds these cell walls will undergo a transition towards a genuine subgrain structure. This occurs through a gradual elimination of extraneous dislocations and the rearrangement of the remaining dislocations into low-angle grain boundaries.

Sub-grain formation is followed by subgrain coarsening where the average size increases while the number of subgrains decreases. This reduces the total area of grain boundary and hence the stored energy in the material. Subgrain coarsen shares many features with grain growth.

If the sub-structure can be approximated to an array of spherical subgrains of radius R and boundary energy $\gamma_s$; the stored energy is uniform; and the force on the boundary is evenly distributed, the driving pressure P is given by:

$$P = -\alpha\, R \frac{d}{dR}\left(\frac{\gamma_s}{R}\right)$$

Since $\gamma_s$ is dependent on the boundary misorientation of the surrounding subgrains, the driving pressure generally does not remain constant throughout coarsening.

## Roasting (Metallurgy)

Roasting is a step of the processing of certain ores. More specifically, roasting is a metallurgical process involving gas–solid reactions at elevated temperatures with the goal of purifying the metal component(s). Often before roasting, the ore has already been partially purified, e.g. by froth floatation. The concentrate is mixed with other materials to facilitate the process. The technology is useful but is also a serious source of air pollution.

Roasted gold ore from Cripple Creek, Colorado. Roasting has driven off the tellurium from the original calaverite, leaving behind vesicular blebs of native gold.

Roasting consists of thermal gas–solid reactions, which can include oxidation, reduction, chlorination, sulfation, and pyrohydrolysis. In roasting, the ore or ore concentrate is treated with very hot air. This process is generally applied to sulfide minerals. During roasting, the sulfide is converted to an oxide, and sulfur is released as sulfur dioxide, a gas. For the ores $Cu_2S$ (chalcocite) and $ZnS$ (sphalerite), balanced equations for the roasting are:

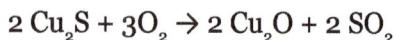

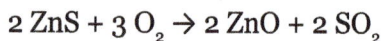

$$2\,Cu_2S + 3O_2 \rightarrow 2\,Cu_2O + 2\,SO_2$$

$$2\,ZnS + 3\,O_2 \rightarrow 2\,ZnO + 2\,SO_2$$

The gaseous product of sulfide roasting, sulfur dioxide ($SO_2$) is often used to produce sulfuric acid. Many sulfide minerals contain other components such as arsenic that are released into the environment.

Up until the early 20th century, roasting was started by burning wood on top of ore. This would raise the temperature of the ore to the point where its sulfur content would become its source of fuel, and the roasting process could continue without external fuel sources. Early sulfide roasting was practiced in this manner in "open hearth" roasters, which were manually stirred (a practice referred to as "rabbling") using rake-like tools to expose unroasted ore to oxygen as the reaction proceeded.

This process would release large amounts of acidic, metallic, and other toxic compounds. Results of this include areas that even after 60-80 years are still largely lifeless, often exactly corresponding to the area of the roast bed, some of which are hundreds of metres wide by kilometres long.

## Roasting Operations

### Oxidizing Roasting

Oxidizing roasting, the most commonly practiced roasting process, involves heating the ore in excess of air or oxygen, to burn out or replace the impurity element, generally sulfur, partly or completely by oxygen. For sulfide roasting, the general reaction can be given by -

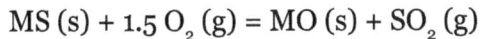

$$MS\ (s) + 1.5\ O_2\ (g) = MO\ (s) + SO_2\ (g)$$

Roasting the sulfide ore, till almost complete removal of the sulfur from the ore, results in a *dead roast*.

## Volatilizing Roasting

Volatilizing roasting, involves careful oxidation at elevated temperatures of the ores, to eliminate impurity elements in the form of their volatile oxides. Examples of such volatile oxides include $As_2O_3$, $Sb_2O_3$, ZnO and sulfur oxides. Careful control of the oxygen content in the roaster is necessary, as excessive oxidation forms non volatile oxides.

## Chloridizing Roasting

Chloridizing roasting transforms certain metal compounds to chlorides, through oxidation or reduction. Some metals like uranium, titanium, beryllium and some rare earths are processed in their chloride form. Certain forms of chloridizing roasting maybe represented by the overall reactions -

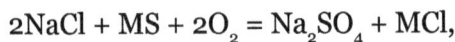

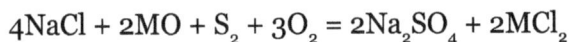

$$2NaCl + MS + 2O_2 = Na_2SO_4 + MCl,$$

$$4NaCl + 2MO + S_2 + 3O_2 = 2Na_2SO_4 + 2MCl_2$$

The first reaction represents the chlorination of a sulfide ore involving an exothermic reaction. The second reaction involving an oxide ore is facilitated by addition of elemental sulfur. Carbonate ores react in a similar manner as the oxide ore, after decomposing to their oxide form at high temperature.

## Sulfating Roasting

Sulfating roasting oxidizes certain sulfide ores to sulfates in a controlled supply of air to enable leaching of the sulfate for further processing.

## Magnetic Roasting

Magnetic roasting involves controlled roasting of the ore to convert it into a magnetic form, thus enabling easy separation and processing in subsequent steps. For example, controlled reduction of haematite (non magnetic $Fe_2O_3$) to magnetite (magnetic $Fe_3O_4$).

## Reduction Roasting

Reduction roasting partially reduces an oxide ore before the actual smelting process.

## Sinter Roasting

Sinter roasting involves heating the fine ores at high temperatures, where simultaneous oxidation and agglomeration of the ores take place. For example, lead sulfide ores are subjected to sinter roasting in a continuous process after froth flotation to convert the fine ores to workable agglomerates for further smelting operations.

# Molding (Process)

Molding or moulding is the process of manufacturing by shaping liquid or pliable raw material using a rigid frame called a mold or matrix. This itself may have been made using a pattern or model of the final object.

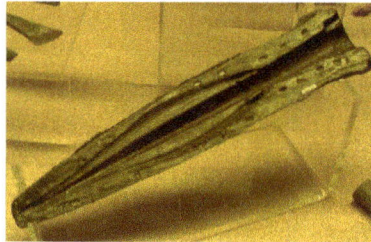

One half of a bronze mold for casting a socketed spear head dated to the period 1400-1000 BC. There are no known parallels for this mold.

Stone mold of the Bronze Age used to produce spear tips.

Ancient Greek molds, used to mass-produce clay figurines, 5th/4th century BC. Beside them, the modern casts taken from them. On display in the Ancient Agora Museum in Athens, housed in the Stoa of Attalus.

Ancient wooden molds used for jaggery & sweets, archaeological museum in Jaffna, Sri Lanka.

A mold or mould is a hollowed-out block that is filled with a liquid or pliable material like plastic, glass, metal, or ceramic raw materials. The liquid hardens or sets inside the mold, adopting its shape.

A mold is the counterpart to a cast. The very common bi-valve molding process uses two molds, one for each half of the object. Piece-molding uses a number of different molds, each creating a section of a complicated object. This is generally only used for larger and more valuable objects.

The manufacturer who makes the molds is called the moldmaker. A release agent is typically used to make removal of the hardened/set substance from the mold easier. Typical uses for molded plastics include molded furniture, molded household goods, molded cases, and structural materials.

Types of molding include:

- Blow molding
- Powder metallurgy plus sintering
- Compression molding
- Extrusion molding
- Injection molding
- Laminating
    - Reaction injection molding
- Matrix molding
- Rotational molding (or Rotomolding)
- Spin casting
- Transfer molding
- Thermoforming
    - Vacuum forming, a simplified version of thermoforming

# Electrowinning

Electrorefining technology converting spent commercial nuclear fuel into metal.

Electrowinning, also called electroextraction, is the electrodeposition of metals from their ores that have been put in solution via a process commonly referred to as leaching. Electrorefining uses a similar process to remove impurities from a metal. Both processes use electroplating on a large scale and are important techniques for the economical and straightforward purification of non-ferrous metals. The resulting metals are said to be *electrowon*.

In electrowinning, a current is passed from an inert anode through a liquid *leach* solution containing the metal so that the metal is extracted as it is deposited in an electroplating process onto the cathode. In electrorefining, the anodes consist of unrefined impure metal, and as the current passes through the acidic electrolyte the anodes are corroded into the solution so that the electroplating process deposits refined pure metal onto the cathodes.

### History

Electrowinning is the oldest industrial electrolytic process. The English chemist Humphry Davy obtained sodium metal in elemental form for the first time in 1807 by the electrolysis of molten sodium hydroxide.

Electrorefining of copper was first demonstrated experimentally by Maximilian, Duke of Leuchtenberg in 1847.

James Elkington patented the commercial process in 1865 and opened the first successful plant in Pembrey, Wales in 1870. The first commercial plant in the United States was the Balbach and Sons Refining and Smelting Company in Newark, New Jersey in 1883.

### Applications

The most common electrowon metals are lead, copper, gold, silver, zinc, aluminium, chromium, cobalt, manganese, and the rare-earth and alkali metals. For aluminium, this is the only production process employed. Several industrially important active metals (which react strongly with water) are produced commercially by electrolysis of their pyrochemical molten salts. Experiments using electrorefining to process spent nuclear fuel have been carried out. Electrorefining may be able to separate heavy metals such as plutonium, caesium, and strontium from the less-toxic bulk of uranium. Many electroextraction systems are also available to remove toxic (and sometimes valuable) metals from industrial waste streams.

### Process

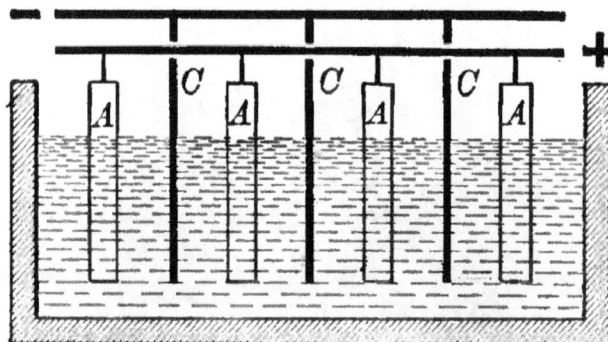

Apparatus for electrolytic refining of copper

Most metals occur in nature in their oxidized form (ores) and thus must be reduced to their metallic forms. The ore is dissolved following some preprocessing in an aqueous electrolyte or in a molten salt and the resulting solution is electrolyzed. The metal is deposited on the cathode (either in solid or in liquid form), while the anodic reaction is usually oxygen evolution. Several metals are naturally present as metal sulfides; these include copper, lead, molybdenum, cadmium, nickel, silver, cobalt, and zinc. In addition, gold and platinum group metals are associated with sulfidic base metal ores. Most metal sulfides or their salts, are electrically conductive and this allows electrochemical redox reactions to efficiently occur in the molten state or in aqueous solutions.

Some metals, such as nickel do not electrolyze out but remain in the electrolyte solution. These are then reduced by chemical reactions to refine the metal. Other metals, which during the processing of the target metal have been reduced but not deposited at the cathode, sink to the bottom of the electrolytic cell, where they form a substance referred to as *anode sludge* or *anode slime*. The metals in this sludge can be removed by standard pyrorefining methods.

Because metal deposition rates are related to available surface area, maintaining properly working cathodes is important. Two cathode types exist, flat-plate and reticulated cathodes, each with its own advantages. Flat-plate cathodes can be cleaned and reused, and plated metals recovered. Reticulated cathodes have a much higher deposition rate compared to flat-plate cathodes. However, they are not reusable and must be sent off for recycling. Alternatively, starter cathodes of pre-refined metal can be used, which become an integral part of the finished metal ready for rolling or further processing.

# Electroplating

Electroplating is a process that uses electric current to reduce dissolved metal cations so that they form a coherent metal coating on an electrode. The term is also used for electrical oxidation of anions onto a solid substrate, as in the formation silver chloride on silver wire to make silver/silver-chloride electrodes. Electroplating is primarily used to change the surface properties of an object (e.g. abrasion and wear resistance, corrosion protection, lubricity, aesthetic qualities, etc.), but may also be used to build up thickness on undersized parts or to form objects by electroforming.

Copper electroplating machine for layering PCBs

The process used in electroplating is called electrodeposition. It is analogous to a galvanic cell acting in reverse. The part to be plated is the cathode of the circuit. In one technique, the anode is made of the metal to be plated on the part. Both components are immersed in a solution called an electrolyte containing one or more dissolved metal salts as well as other ions that permit the flow of electricity. A power supply supplies a direct current to the anode, oxidizing the metal atoms that it comprises and allowing them to dissolve in the solution. At the cathode, the dissolved metal ions in the electrolyte solution are reduced at the interface between the solution and the cathode, such that they "plate out" onto the cathode. The rate at which the anode is dissolved is equal to the rate at which the cathode is plated, vis-a-vis the current through the circuit. In this manner, the ions in the electrolyte bath are continuously replenished by the anode.

Other electroplating processes may use a non-consumable anode such as lead or carbon. In these techniques, ions of the metal to be plated must be periodically replenished in the bath as they are drawn out of the solution. The most common form of electroplating is used for creating coins such as pennies, which are small zinc plates covered in a layer of copper.

**Process**

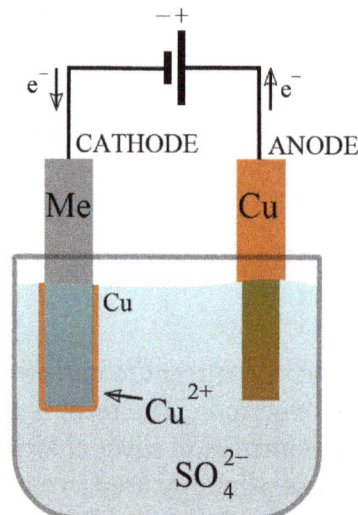

Electroplating of a metal (Me) with copper in a copper sulfate bath

The cations associate with the anions in the solution. These cations are reduced at the cathode to deposit in the metallic, zero valence state. For example, in an acid solution, copper is oxidized at the anode to $Cu^{2+}$ by losing two electrons. The $Cu^{2+}$ associates with the anion $SO_4^{2-}$ in the solution to form copper sulfate. At the cathode, the $Cu^{2+}$ is reduced to metallic copper by gaining two electrons. The result is the effective transfer of copper from the anode source to a plate covering the cathode.

The plating is most commonly a single metallic element, not an alloy. However, some alloys can be electrodeposited, notably brass and solder.

Many plating baths include cyanides of other metals (e.g., potassium cyanide) in addition to cyanides of the metal to be deposited. These free cyanides facilitate anode corrosion, help to maintain a constant metal ion level and contribute to conductivity. Additionally, non-metal chemicals such as carbonates and phosphates may be added to increase conductivity.

When plating is not desired on certain areas of the substrate, stop-offs are applied to prevent the bath from coming in contact with the substrate. Typical stop-offs include tape, foil, lacquers, and waxes.

The ability of a plating to cover uniformly is called *throwing power*; the better the "throwing power" the more uniform the coating.

## Strike

Initially, a special plating deposit called a "strike" or "flash" may be used to form a very thin (typically less than 0.1 micrometer thick) plating with high quality and good adherence to the substrate. This serves as a foundation for subsequent plating processes. A strike uses a high current density and a bath with a low ion concentration. The process is slow, so more efficient plating processes are used once the desired strike thickness is obtained.

The striking method is also used in combination with the plating of different metals. If it is desirable to plate one type of deposit onto a metal to improve corrosion resistance but this metal has inherently poor adhesion to the substrate, a strike can be first deposited that is compatible with both. One example of this situation is the poor adhesion of electrolytic nickel on zinc alloys, in which case a copper strike is used, which has good adherence to both.

## Electrochemical Deposition

Electrochemical deposition is generally used for the growth of metals and conducting metal oxides because of the following advantages: (i) the thickness and morphology of the nanostructure can be precisely controlled by adjusting the electrochemical parameters, (ii) relatively uniform and compact deposits can be synthesized in template-based structures, (iii) higher deposition rates are obtained, and (iv) the equipment is inexpensive due to the non-requirements of either a high vacuum or a high reaction temperature.

## Pulse Electroplating or Pulse Electrodeposition (PED)

A simple modification in the electroplating process is the pulse electroplating. This process involves the swift alternating of the potential or current between two different values resulting in a series of pulses of equal amplitude, duration and polarity, separated by zero current. By changing the pulse amplitude and width, it is possible to change the deposited film's composition and thickness.

## Brush Electroplating

A closely related process is brush electroplating, in which localized areas or entire items are plated using a brush saturated with plating solution. The brush, typically a stainless steel body wrapped with a cloth material that both holds the plating solution and prevents direct contact with the item being plated, is connected to the positive side of a low voltage direct-current power source, and the item to be plated connected to the negative. The operator dips the brush in plating solution then applies it to the item, moving the brush continually to get an even distribution of the plating material. Brush electroplating has several advantages over tank plating, including portability, ability to plate items that for some reason cannot be tank plated (one application was the plating of portions of very large decorative support columns in a building restoration), low or no masking requirements, and comparatively low plating solution volume requirements. Disadvantages compared to tank plating can include greater operator involvement (tank plating can frequently be done with minimal attention), and inability to achieve as great a plate thickness.

## Electroless Deposition

Usually an electrolytic cell (consisting of two electrodes, electrolyte, and external source of current) is used for electrodeposition. In contrast, an electroless deposition process uses only one electrode and no external source of electric current. However, the solution for the electroless process needs to contain a reducing agent so that the electrode reaction has the form:

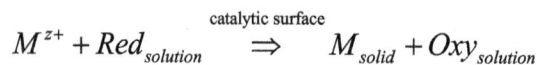

$$M^{z+} + Red_{solution} \xrightarrow{\text{catalytic surface}} M_{solid} + Oxy_{solution}$$

In principle any hydrogen-based reducer can be used although the redox potential of the reducer half-cell must be high enough to overcome the energy barriers inherent in liquid chemistry. Electroless nickel plating uses hypophosphite as the reducer while plating of other metals like silver, gold and copper typically use low molecular weight aldehydes.

A major benefit of this approach over electroplating is that the power sources and plating baths are not needed, reducing the cost of production. The technique can also plate diverse shapes and types of surface. The downside is that the plating process is usually slower and cannot create such thick plates of metal. As a consequence of these characteristics, electroless deposition is quite common in the decorative arts.

## Cleanliness

Cleanliness is essential to successful electroplating, since molecular layers of oil can prevent adhesion of the coating. ASTM B322 is a standard guide for cleaning metals prior to electroplating. Cleaning processes include solvent cleaning, hot alkaline detergent cleaning, electro-cleaning, and acid treatment etc. The most common industrial test for cleanliness is the waterbreak test, in which the surface is thoroughly rinsed and held vertical. Hydrophobic contaminants such as oils cause the water to bead and break up, allowing the water to drain rapidly. Perfectly clean metal surfaces are hydrophilic and will retain an unbroken sheet of water that does not bead up or drain off. ASTM F22 describes a version of this test. This test does not detect hydrophilic contaminants, but the electroplating process can displace these easily since the solutions are water-based. Surfactants such as soap reduce the sensitivity of the test and must be thoroughly rinsed off.

## Effects

Electroplating changes the chemical, physical, and mechanical properties of the workpiece. An example of a chemical change is when nickel plating improves corrosion resistance. An example of a physical change is a change in the outward appearance. An example of a mechanical change is a change in tensile strength or surface hardness which is a required attribute in tooling industry. Electroplating of acid gold on underlying copper/nickel-plated circuits reduces contact resistance as well as surface hardness. Copper-plated areas of mild steel act as a mask if case hardening of such areas are not desired. Tin-plated steel is chromium-plated to prevent dulling of the surface due to oxidation of tin.

## History

Modern electrochemistry was invented by Italian chemist Luigi Valentino Brugnatelli (it) in 1805. Brugnatelli used his colleague Alessandro Volta's invention of five years earlier, the voltaic pile, to facilitate the first electrodeposition. Brugnatelli's inventions were suppressed by the French Academy of Sciences and did not become used in general industry for the following thirty years.

Luigu Valentino Brugnatelli

Nickel plating

By 1839, scientists in Britain and Russia had independently devised metal deposition processes similar to Brugnatelli's for the copper electroplating of printing press plates.

Boris Jacobi developed electroplating, electrotyping and galvanoplastic sculpture in Russia

Boris Jacobi in Russia not only rediscovered galvanoplastics, but developed electrotyping and galvanoplastic sculpture. Galvanoplastics quickly came into fashion in Russia, with such people as inventor Peter Bagration, scientist Heinrich Lenz and science fiction author Vladimir Odoyevsky all contributing to further development of the technology. Among the most notorious cases of electroplating usage in mid-19th century Russia were gigantic galvanoplastic sculptures of St. Isaac's Cathedral in Saint Petersburg and gold-electroplated dome of the Cathedral of Christ the Saviour in Moscow, the tallest Orthodox church in the world.

Galvanoplastic sculpture on St. Isaac's Cathedral in Saint Petersburg.

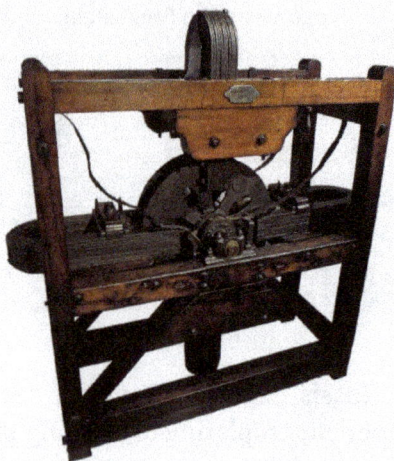

The Woolrich Electrical Generator in Thinktank, Birmingham

Soon after, John Wright of Birmingham, England discovered that potassium cyanide was a suitable electrolyte for gold and silver electroplating. Wright's associates, George Elkington and Henry Elkington were awarded the first patents for electroplating in 1840. These two then founded the electroplating industry in Birmingham from where it spread around the world. The Woolrich Electrical Generator of 1844, now in Thinktank, Birmingham Science Museum, is the earliest electrical generator used in an industrial process. It was used by Elkingtons.

The Norddeutsche Affinerie in Hamburg was the first modern electroplating plant starting its production in 1876.

As the science of electrochemistry grew, its relationship to the electroplating process became understood and other types of non-decorative metal electroplating processes were developed. Commercial electroplating of nickel, brass, tin, and zinc were developed by the 1850s. Electroplating baths and equipment based on the patents of the Elkingtons were scaled up to accommodate the plating of numerous large scale objects and for specific manufacturing and engineering applications.

The plating industry received a big boost with the advent of the development of electric generators in the late 19th century. With the higher currents available, metal machine components, hardware, and automotive parts requiring corrosion protection and enhanced wear properties, along with better appearance, could be processed in bulk.

The two World Wars and the growing aviation industry gave impetus to further developments and refinements including such processes as hard chromium plating, bronze alloy plating, sulfamate nickel plating, along with numerous other plating processes. Plating equipment evolved from manually operated tar-lined wooden tanks to automated equipment, capable of processing thousands of kilograms per hour of parts.

One of the American physicist Richard Feynman's first projects was to develop technology for electroplating metal onto plastic. Feynman developed the original idea of his friend into a successful invention, allowing his employer (and friend) to keep commercial promises he had made but could not have fulfilled otherwise.

## Uses

Electroplating is widely used in various industries for coating metal objects with a thin layer of a different metal. The layer of metal deposited has some desired property, which the metal of the object lacks. For example, chromium plating is done on many objects such as car parts, bath taps, kitchen gas burners, wheel rims and many others for the fact that chromium is very corrosion resistant, and thus prolongs the life of the parts. Electroplating has wide usage in industries. It is also used in making inexpensive jewelry. Electroplating increases life of metal and prevents corrosion.

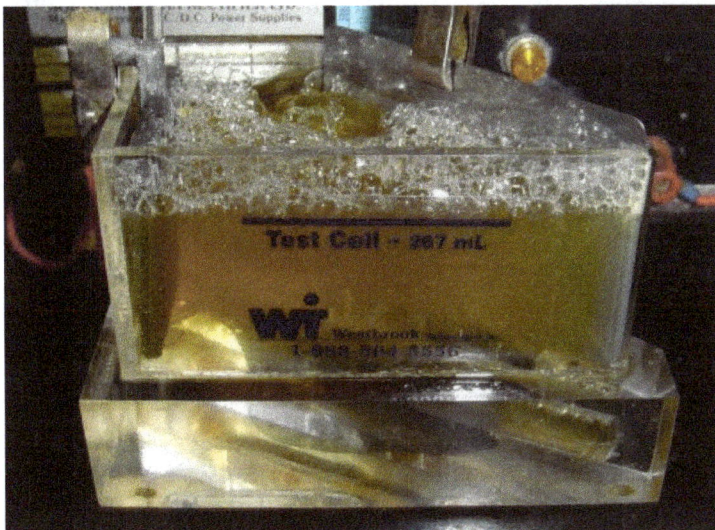

A zinc solution tested in a hull cell

## Hull Cell

The *Hull cell* is a type of test cell used to qualitatively check the condition of an electroplating bath. It allows for optimization for current density range, optimization of additive concentration, recognition of impurity effects and indication of macro-throwing power capability. The Hull cell replicates the plating bath on a lab scale. It is filled with a sample of the plating solution, an appropriate anode which is connected to a rectifier. The "work" is replaced with a hull cell test panel that will be plated to show the "health" of the bath.

The Hull cell is a trapezoidal container that holds 267 ml of solution. This shape allows one to place the test panel on an angle to the anode. As a result, the deposit is plated at different current densities which can be measured with a hull cell ruler. The solution volume allows for a quantitative optimization of additive concentration: 1 gram addition to 267 mL is equivalent to 0.5 oz/gal in the plating tank.

## Haring-Blum Cell

The Haring-Blum Cell is used to determine the macro throwing power of a plating bath. The cell consists of 2 parallel cathodes with a fixed anode in the middle. The cathodes are at distances from the anode in the ratio of 1:5. The macro throwing power is calculating from the thickness of plating at the two cathodes when a dc current is passed for a specific period of time. The cell is fabricated out of Perspex or glass.

## References

- Degarmo, E. Paul; Black, J T.; Kohser, Ronald A. (2003), Materials and Processes in Manufacturing (9th ed.), Wiley, p. 277, ISBN 0-471-65653-4
- Gordon Elliott (2006). Aspects of Ceramic History: A Series of Papers Focusing on the Ceramic Artifact As Evidence of Cultural and Technical Developments. Gordon Elliott. pp. 52–. ISBN 978-0-9557690-0-9.
- B. RAVI (1 January 2005). METAL CASTING: COMPUTER-AIDED DESIGN AND ANALYSIS. PHI Learning Pvt. Ltd. pp. 92–. ISBN 978-81-203-2726-9.
- T F Waters (11 September 2002). Fundamentals of Manufacturing For Engineers. CRC Press. pp. 17–. ISBN 978-0-203-50018-7.
- James T. Frane (1994). Craftsman's Illustrated Dictionary of Construction Terms. Craftsman Book Company. pp. 126–. ISBN 978-1-57218-008-6.
- Jane L. Bassett; Peggy Fogelman; David A. Scott; Ronald C. Schmidtling (2008). The Craftsman Revealed: Adriaen de Vries, Sculptor in Bronze. Getty Publications. pp. 269–. ISBN 978-0-89236-919-5.
- Atkins, Peter; Jones, Loretta (2008), Chemical Principles: The Quest for Insight (4th ed.), W. H. Freeman and Company, p. 236, ISBN 0-7167-7355-4
- Ott, J. Bevan; Boerio-Goates, Juliana (2000), Chemical Thermodynamics: Advanced Applications, Academic Press, pp. 92–93, ISBN 0-12-530985-6
- Greenwood, Norman N.; Earnshaw, Alan (1997). Chemistry of the Elements (2nd ed.). Butterworth-Heinemann. ISBN 0-08-037941-9.
- Ray, H.S.; et al. (1985). Extraction of Nonferrous Metals. Affiliated East-West Press Private Limited. pp. 131,132. ISBN 81-85095-63-9.
- Todd, Robert H.; Dell K. Allen; Leo Alting (1994). "Surface Coating". Manufacturing Processes Reference

Guide. Industrial Press Inc. pp. 454–458. ISBN 0-8311-3049-0

- Thomas, John Meurig (1991). Michael Faraday and the Royal Institution: The Genius of Man and Place. Bristol: Hilger. p. 51. ISBN 0750301457.

- Bard, Allan; Inzelt, Gyorgy; Scholz, Fritz, eds. (2012). "Haring-Blum Cell". Electrochemical Dictionary. Springer. p. 444. doi:10.1007/978-3-642-29551-5_8. ISBN 978-3-642-29551-5.

- Wendt, Hartmut; Kreyse, Gerhard (1999). Electrochemical Engineering: Science and Technology in Chemical and Other Industries. Springer. p. 122. ISBN 3540643869

- "Mining, Quarrying & Prospecting: The Difference between Mining, Quarrying & Prospecting". mqp-geotek. blogspot.co.uk. Retrieved 2015-06-11.

- "Assessing the opportunities of landfill mining - Research database - University of Groningen". www.rug.nl. Retrieved 2015-06-11.

- Cambell, Bonnie (2008). "Regulation & Legitimacy in the Mining Industry in Africa: Where does" (PDF). Review of African Political Economy 35 (3): 367–389. doi:10.1080/03056240802410984. Retrieved 7 April 2011.

- The World Bank. ces.worldbank.org/INTOGMC/Resources/336099-1288881181404/7530465-1288881207444/ eifd19_mining_sector_reform.pdf "The World Bank's Evolutionary Approach to Mining Sector Reform" Check |url= value (help) (PDF). Retrieved 4 April 2011.

# 4

# Alloy: An Integrated Study

This chapter studies the various metal alloys in great detail and also has a section dedicated to non-ferrous metals. Alloys are manufactured as they provide better corrosion resistance as well as mechanical as well as torsion strength. Readers are provided with information regarding the formation and uses of the alloys.

## Alloy

Wire rope made from steel, which is a metal alloy whose major component is iron, with carbon content between 0.02% and 2.14% by mass.

An alloy is a mixture of metals or a mixture of a metal and another element. Alloys are defined by metallic bonding character. An alloy may be a solid solution of metal elements (a single phase) or a mixture of metallic phases (two or more solutions). Intermetallic compounds are alloys with a defined stoichiometry and crystal structure. Zintl phases are also sometimes considered alloys depending on bond types.

Alloys are used in a wide variety of applications. In some cases, a combination of metals may reduce the overall cost of the material while preserving important properties. In other cases, the combination of metals imparts synergistic properties to the constituent metal elements such as corrosion resistance or mechanical strength. Examples of alloys are steel, solder, brass, pewter, duralumin, phosphor bronze and amalgams.

The alloy constituents are usually measured by mass. Alloys are usually classified as substitutional or interstitial alloys, depending on the atomic arrangement that forms the alloy. They can be further classified as homogeneous (consisting of a single phase), or heterogeneous (consisting of two or more phases) or intermetallic.

## Introduction

An alloy is a mixture of either pure or fairly pure chemical elements, which forms an impure substance (admixture) that retains the characteristics of a metal. An alloy is distinct from an impure metal, such as wrought iron, in that, with an alloy, the added impurities are usually desirable and will typically have some useful benefit. Alloys are made by mixing two or more elements; at least one of which being a metal. This is usually called the primary metal or the base metal, and the name of this metal may also be the name of the alloy. The other constituents may or may not be metals but, when mixed with the molten base, they will be soluble, dissolving into the mixture.

Liquid bronze, being poured into molds during casting.

A brass lamp.

When the alloy cools and solidifies (crystallizes), its mechanical properties will often be quite different from those of its individual constituents. A metal that is normally very soft and malleable, such as aluminium, can be altered by alloying it with another soft metal, like copper. Although both metals are very soft and ductile, the resulting aluminium alloy will be much harder and stronger. Adding a small amount of non-metallic carbon to iron produces an alloy called steel. Due to its very-high strength and toughness (which is much higher than pure iron), and its ability to be greatly altered by heat treatment, steel is one of the most common alloys in modern use. By adding chromium to steel, its resistance to corrosion can be enhanced, creating stainless steel, while adding silicon will alter its electrical characteristics, producing silicon steel.

Although the elements usually must be soluble in the liquid state, they may not always be soluble in the solid state. If the metals remain soluble when solid, the alloy forms a solid solution, becoming a homogeneous structure consisting of identical crystals, called a phase. If the mixture cools and the constituents become insoluble, they may separate to form two or more different types of crystals, creating a heterogeneous microstructure of different phases. However, in other alloys, the insoluble elements may not separate until after crystallization occurs. These alloys are called intermetallic alloys because, if cooled very quickly, they first crystallize as a homogeneous phase, but they are supersaturated with the secondary constituents. As time passes, the atoms of these supersaturated alloys separate within the crystals, forming intermetallic phases that serve to reinforce the crystals internally.

Some alloys occur naturally, such as electrum, which is an alloy that is native to Earth, consisting of silver and gold. Meteorites are sometimes made of naturally occurring alloys of iron and nickel, but are not native to the Earth. One of the first alloys made by humans was bronze, which is made by mixing the metals tin and copper. Bronze was an extremely useful alloy to the ancients, because it is much stronger and harder than either of its components. Steel was another common alloy. However, in ancient times, it could only be created as an accidental byproduct from the heating of iron ore in fires (smelting) during the manufacture of iron. Other ancient alloys include pewter, brass and pig iron. In the modern age, steel can be created in many forms. Carbon steel can be made by varying only the carbon content, producing soft alloys like mild steel or hard alloys like spring steel. Alloy steels can be made by adding other elements, such as molybdenum, vanadium or nickel, resulting in alloys such as high-speed steel or tool steel. Small amounts of manganese are usually alloyed with most modern-steels because of its ability to remove unwanted impurities, like phosphorus, sulfur and oxygen, which can have detrimental effects on the alloy. However, most alloys were not created until the 1900s, such as various aluminium, titanium, nickel, and magnesium alloys. Some modern superalloys, such as incoloy, inconel, and hastelloy, may consist of a multitude of different components.

## Terminology

The term alloy is used to describe a mixture of atoms in which the primary constituent is a metal. The primary metal is called the *base*, the *matrix*, or the *solvent*. The secondary constituents are often called *solutes*. If there is a mixture of only two types of atoms, not counting impurities, such as a copper-nickel alloy, then it is called a *binary alloy*. If there are three types of atoms forming the mixture, such as iron, nickel and chromium, then it is called a *ternary alloy*. An alloy with four constituents is a *quaternary alloy,* while a five-part alloy is termed a *quinary alloy*. Because the percentage of each constituent can be varied, with any mixture the entire range of possible variations is called a *system*. In this respect, all of the various forms of an alloy containing only two constituents, like iron and carbon, is called a *binary system,* while all of the alloy combinations possible with a ternary alloy, such as alloys of iron, carbon and chromium, is called a *ternary system.*

A gate valve, made from Inconel.

Although an alloy is technically an impure metal, when referring to alloys, the term "impurities" usually denotes those elements which are not desired. These impurities are often found in the base metals or the solutes, but they may also be introduced during the alloying process. For instance, sulfur is a common impurity in steel. Sulfur combines readily with iron to form iron sulfide, which is very brittle, creating weak spots in the steel. Lithium, sodium and calcium are common impurities in aluminium alloys, which can have adverse effects on the structural integrity of castings. Conversely, otherwise pure-metals that simply contain unwanted impurities are often called "impure metals" and are not usually referred to as alloys. Oxygen, present in the air, readily combines with most metals to form metal oxides; especially at higher temperatures encountered during alloying. Great care is often taken during the alloying process to remove excess impurities, using fluxes, chemical additives, or other methods of extractive metallurgy.

In practice, some alloys are used so predominantly with respect to their base metals that the name of the primary constituent is also used as the name of the alloy. For example, 14 karat gold is an alloy of gold with other elements. Similarly, the silver used in jewelry and the aluminium used as a structural building material are also alloys.

The term "alloy" is sometimes used in everyday speech as a synonym for a particular alloy. For example, automobile wheels made of an aluminium alloy are commonly referred to as simply "alloy wheels", although in point of fact steels and most other metals in practical use are also alloys. Steel is such a common alloy that many items made from it, like wheels, barrels, or girders, are simply referred to by the name of the item, assuming it is made of steel. When made from other materials, they are typically specified as such, (i.e.: "bronze wheel", "plastic barrel", or "wood girder").

## Theory

Alloying a metal is done by combining it with one or more other metals or non-metals that often enhance its properties. For example, steel is stronger than iron, its primary element. The electrical and thermal conductivity of alloys is usually lower than that of the pure metals. The physical properties, such as density, reactivity, Young's modulus of an alloy may not differ greatly from those of its elements, but engineering properties such as tensile strength and shear strength may be substantially different from those of the constituent materials. This is sometimes a result of the sizes of the atoms in the alloy, because larger atoms exert a compressive force on neighboring atoms, and smaller atoms exert a tensile force on their neighbors, helping the alloy resist deformation. Sometimes alloys may exhibit marked differences in behavior even when small amounts of one element are present. For example, impurities in semiconducting ferromagnetic alloys lead to different properties, as first predicted by White, Hogan, Suhl, Tian Abrie and Nakamura. Some alloys are made by melting and mixing two or more metals. Bronze, an alloy of copper and tin, was the first alloy discovered, during the prehistoric period now known as the bronze age; it was harder than pure copper and originally used to make tools and weapons, but was later superseded by metals and alloys with better properties. In later times bronze has been used for ornaments, bells, statues, and bearings. Brass is an alloy made from copper and zinc.

Unlike pure metals, most alloys do not have a single melting point, but a melting range in which the material is a mixture of solid and liquid phases. The temperature at which melting begins is called the solidus, and the temperature when melting is just complete is called the liquidus. However, for many alloys there is a particular proportion of constituents (in some cases more than one)—either a eutectic mixture or a peritectic composition—which gives the alloy a unique melting point.

## Heat-treatable Alloys

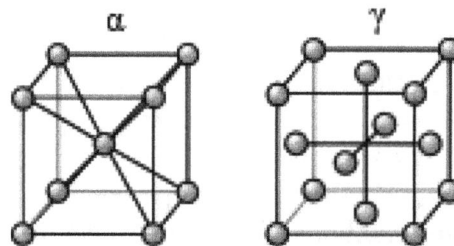

Allotropes of iron, (alpha iron and gamma iron) showing the differences in atomic arrangement.

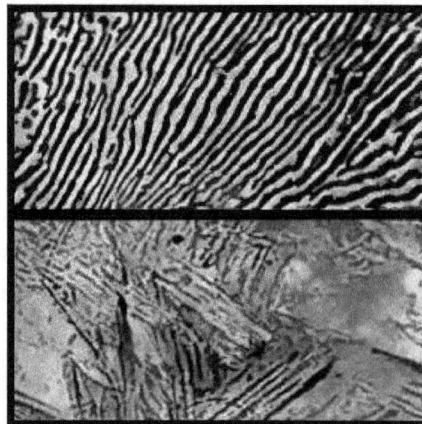

Photomicrographs of steel. Top photo: Annealed (slowly cooled) steel forms a heterogeneous, lamellar microstructure called pearlite, consisting of the phases cementite (light) and ferrite (dark). Bottom photo: Quenched (quickly cooled) steel forms a single phase called martensite, in which the carbon remains trapped within the crystals, creating internal stresses.

Alloys are often made to alter the mechanical properties of the base metal, to induce hardness, toughness, ductility, or other desired properties. Most metals and alloys can be work hardened by creating defects in their crystal structure. These defects are created during plastic deformation, such as hammering or bending, and are permanent unless the metal is recrystallized. However, some alloys can also have their properties altered by heat treatment. Nearly all metals can be softened by annealing, which recrystallizes the alloy and repairs the defects, but not as many can be hardened by controlled heating and cooling. Many alloys of aluminium, copper, magnesium, titanium, and nickel can be strengthened to some degree by some method of heat treatment, but few respond to this to the same degree that steel does.

At a certain temperature, (usually between 1,500 °F (820 °C) and 1,600 °F (870 °C), depending on carbon content), the base metal of steel undergoes a change in the arrangement of the atoms in its crystal matrix, called allotropy. This allows the small carbon atoms to enter the interstices of the iron crystal, diffusing into the iron matrix. When this happens, the carbon atoms are said to be in *solution,* or mixed with the iron, forming a single, homogeneous, crystalline phase called austenite. If the steel is cooled slowly, the iron will gradually change into its low temperature allotrope. When this happens the carbon atoms will no longer be soluble with the iron, and will be forced to precipitate out of solution, nucleating into the spaces between the crystals. The steel then becomes heterogeneous, being formed of two phases; the carbon (carbide) phase cementite, and ferrite. This type of heat treatment produces steel that is rather soft and bendable. However, if the steel is cooled quickly the carbon atoms will not have time to precipitate. When rapidly cooled, a diffusionless (martensite) transformation occurs, in which the carbon atoms become trapped in solution. This causes the iron crystals to deform intrinsically when the crystal structure tries to change to its low temperature state, making it very hard and brittle.

Conversely, most heat-treatable alloys are precipitation hardening alloys, which produce the opposite effects that steel does. When heated to form a solution and then cooled quickly, these alloys become much softer than normal, during the diffusionless transformation, and then harden as they age. The solutes in these alloys will precipitate over time, forming intermetallic phases, which are difficult to discern from the base metal. Unlike steel, in which the solid solution separates to form different crystal phases, precipitation hardening alloys separate to form different phases within the same crystal. These intermetallic alloys appear homogeneous in crystal structure, but tend to behave heterogeneous, becoming hard and somewhat brittle.

## Substitutional and Interstitial Alloys

When a molten metal is mixed with another substance, there are two mechanisms that can cause an alloy to form, called *atom exchange* and the *interstitial mechanism*. The relative size of each element in the mix plays a primary role in determining which mechanism will occur. When the atoms are relatively similar in size, the atom exchange method usually happens, where some of the atoms composing the metallic crystals are substituted with atoms of the other constituent. This is called a *substitutional alloy.* Examples of substitutional alloys include bronze and brass, in which some of the copper atoms are substituted with either tin or zinc atoms. With the interstitial mechanism, one atom is usually much smaller than the other, so cannot successfully replace an atom in the crystals of the base metal. The smaller atoms become trapped in the spaces between the atoms in the crystal matrix, called the *interstices*. This is referred to as an *interstitial alloy*.

Steel is an example of an interstitial alloy, because the very small carbon atoms fit into interstices of the iron matrix. Stainless steel is an example of a combination of interstitial and substitutional alloys, because the carbon atoms fit into the interstices, but some of the iron atoms are replaced with nickel and chromium atoms.

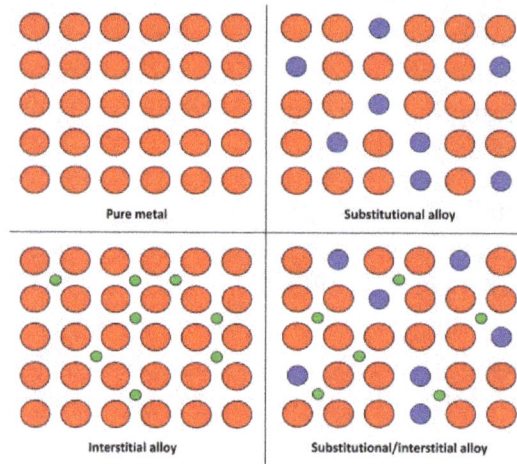

Different atomic mechanisms of alloy formation, showing pure metal, substitutional, interstitial, and a combination of the two.

## History and Examples

### Meteoric Iron

A meteorite and a hatchet that was forged from meteoric iron.

The use of alloys by humans started with the use of meteoric iron, a naturally occurring alloy of nickel and iron. It is the main constituent of iron meteorites which occasionally fall down on Earth from outer space. As no metallurgic processes were used to separate iron from nickel, the alloy was used as it was. Meteoric iron could be forged from a red heat to make objects such as tools, weapons, and nails. In many cultures it was shaped by cold hammering into knives and arrowheads. They were often used as anvils. Meteoric iron was very rare and valuable, and difficult for ancient people to work.

# Bronze and Brass

Bronze axe 1100 BC

Bronze doorknocker

Iron is usually found as iron ore on Earth, except for one deposit of native iron in Greenland, which was used by the Inuit people. Native copper, however, was found worldwide, along with silver, gold and platinum, which were also used to make tools, jewelry, and other objects since Neolithic times. Copper was the hardest of these metals, and the most widely distributed. It became one of the most important metals to the ancients. Eventually, humans learned to smelt metals such as copper and tin from ore, and, around 2500 BC, began alloying the two metals to form bronze, which is much harder than its ingredients. Tin was rare, however, being found mostly in Great Britain. In the Middle East, people began alloying copper with zinc to form brass. Ancient civilizations took into account the mixture and the various properties it produced, such as hardness, toughness and melting point, under various conditions of temperature and work hardening, developing much of the information contained in modern alloy phase diagrams. Arrowheads from the Chinese Qin dynasty (around 200 BC) were often constructed with a hard bronze-head, but a softer bronze-tang, combining the alloys to prevent both dulling and breaking during use.

## Amalgams

Mercury has been smelted from cinnabar for thousands of years. Mercury dissolves many metals, such as gold, silver, and tin, to form amalgams (an alloy in a soft paste, or liquid form at ambient temperature). Amalgams have been used since 200 BC in China for plating objects with precious metals, called gilding, such as armor and mirrors. The ancient Romans often used mercury-tin amalgams for gilding their armor. The amalgam was applied as a paste and then heated until the mercury vaporized, leaving the gold, silver, or tin behind. Mercury was often used in mining, to extract precious metals like gold and silver from their ores.

## Precious-Metal Alloys

Electrum, a natural alloy of silver and gold, was often used for making coins.

Many ancient civilizations alloyed metals for purely aesthetic purposes. In ancient Egypt and Mycenae, gold was often alloyed with copper to produce red-gold, or iron to produce a bright burgundy-gold. Gold was often found alloyed with silver or other metals to produce various types of colored gold. These metals were also used to strengthen each other, for more practical purposes. Copper was often added to silver to make sterling silver, increasing its strength for use in dishes, silverware, and other practical items. Quite often, precious metals were alloyed with less valuable substances as a means to deceive buyers. Around 250 BC, Archimedes was commissioned by the king to find a way to check the purity of the gold in a crown, leading to the famous bath-house shouting of "Eureka!" upon the discovery of Archimedes' principle.

## Pewter

The term pewter covers a variety of alloys consisting primarily of tin. As a pure metal, tin was much too soft to be used for any practical purpose. However, in the Bronze age, tin was a rare metal and, in many parts of Europe and the Mediterranean, was often valued higher than gold. To make jewelry, forks and spoons, or other objects from tin, it was usually alloyed with other metals to increase its strength and hardness. These metals were typically lead, antimony, bismuth or copper. These solutes sometimes were added individually in varying amounts, or added together, making a wide variety of things, ranging from practical items, like dishes, surgical tools, candlesticks or funnels, to decorative items such as ear rings and hair clips.

The earliest examples of pewter come from ancient Egypt, around 1450 BC. The use of pewter was widespread across Europe, from France to Norway and Britain (where most of the ancient tin was mined) to the Near East. The alloy was also used in China and the Far East, arriving in Japan around 800 AD, where it was used for making objects like ceremonial vessels, tea canisters, or chalices used in shinto shrines.

## Steel and Pig Iron

The first known smelting of iron began in Anatolia, around 1800 BC. Called the bloomery process, it produced very soft but ductile wrought iron. By 800 BC, iron-making technology had spread to Europe, arriving in Japan around 700 AD. Pig iron, a very hard but brittle alloy of iron and carbon, was being produced in China as early as 1200 BC, but did not arrive in Europe until the Middle Ages. Pig iron has a lower melting point than iron, and was used for making cast-iron. However, these metals found little practical use until the introduction of crucible steel around 300 BC. These steels were of poor quality, and the introduction of pattern welding, around the 1st century AD, sought to balance the extreme properties of the alloys by laminating them, to create a tougher metal. Around 700 AD, the Japanese began folding bloomery-steel and cast-iron in alternating layers to increase the strength of their swords, using clay fluxes to remove slag and impurities. This method of Japanese swordsmithing produced one of the purest steel-alloys of the early Middle Ages.

While the use of iron started to become more widespread around 1200 BC, mainly because of interruptions in the trade routes for tin, the metal is much softer than bronze. However, very small amounts of steel, (an alloy of iron and around 1% carbon), was always a byproduct of the bloomery process. The ability to modify the hardness of steel by heat treatment had been known since 1100 BC, and the rare material was valued for the manufacture of tools and weapons. Because the ancients could not produce temperatures high enough to melt iron fully, the production of steel in decent quantities did not occur until the introduction of blister steel during the Middle Ages. This method introduced carbon by heating wrought iron in charcoal for long periods of time, but the penetration of carbon was not very deep, so the alloy was not homogeneous. In 1740, Benjamin Huntsman began melting blister steel in a crucible to even out the carbon content, creating the first process for the mass production of tool steel. Huntsman's process was used for manufacturing tool steel until the early 1900s.

With the introduction of the blast furnace to Europe in the Middle Ages, pig iron was able to be produced in much higher volumes than wrought iron. Because pig iron could be melted, people began to develop processes of reducing the carbon in the liquid pig iron to create steel. Puddling was introduced during the 1700s, where molten pig iron was stirred while exposed to the air, to remove the carbon by oxidation. In 1858, Sir Henry Bessemer developed a process of steel-making by blowing hot air through liquid pig iron to reduce the carbon content. The Bessemer process was able to produce the first large scale manufacture of steel. Once the Bessemer process began to gain widespread use, other alloys of steel began to follow. Mangalloy, an alloy of steel and manganese exhibiting extreme hardness and toughness, was one of the first alloy steels, and was created by Robert Hadfield in 1882.

## Precipitation-Hardening Alloys

In 1906, precipitation hardening alloys were discovered by Alfred Wilm. Precipitation hardening alloys, such as certain alloys of aluminium, titanium, and copper, are heat-treatable alloys that soften when quenched (cooled quickly), and then harden over time. After quenching a ternary alloy of aluminium, copper, and magnesium, Wilm discovered that the alloy increased in hardness when left to age at room temperature. Although an explanation for the phenomenon was not provided until 1919, duralumin was one of the first "age hardening" alloys to be used, and was soon followed by many others. Because they often exhibit a combination of high strength and low weight, these alloys became widely used in many forms of industry, including the construction of modern aircraft.

## Ferrous Metallurgy

Bloomery smelting during the Middle Ages.

Ferrous metallurgy involves processes and alloys based on iron. It began far back in prehistory. The earliest surviving iron artifacts, from the 4th millennium BC in Egypt, were made from meteoritic iron-nickel. It is not known when or where the smelting of iron from ores began, but by the end of the 2nd millennium BC iron was being produced from iron ores from China to Africa south of the Sahara. The use of wrought iron (worked iron) was known by the 1st millennium BC. During the medieval period, means were found in Europe of producing wrought iron from cast iron (in this context known as pig iron) using finery forges. For all these processes, charcoal was required as fuel.

Steel (with a carbon content between pig iron and wrought iron) was first produced in antiquity as an alloy. Its process of production, Wootz, was exported before the 4th century BC to ancient China, Africa, the Middle East and Europe. Archaeological evidence of cast iron appears in 5th century BC China. New methods of producing it by carburizing bars of iron in the cementation process were devised in the 17th century. During the Industrial Revolution, new methods of producing bar iron without charcoal were devised and these were later applied to produce steel. In the late 1850s, Henry Bessemer invented a new steelmaking process, that involved blowing air through molten pig iron to burn off carbon, and so to produce mild steel. This and other 19th-century and later processes have displaced the use of wrought iron. Today, wrought iron is no longer produced on a commercial scale.

# Meteoric Iron

Willamette Meteorite, the sixth largest in the world, is an iron-nickel meteorite

Iron meteorites consist overwhelmingly of nickel-iron alloys. The metal taken from these meteorites is known as meteoric iron and was one of the earliest sources of usable iron available to humans.

Iron was extracted from iron-nickel meteorites, which comprise about 6% of all meteorites that fall on the earth. That source can often be identified with certainty because of the unique crystalline features ("Widmanstatten figures") of that material, which are preserved when the metal is worked cold or at low temperature. Those artifacts include, for example, a bead from the 5th millennium BC found in Iran and spear tips and ornaments from Ancient Egypt and Sumer around 4000 BC. Meteoric iron has been identified also in a Chinese axe head from the middle of the 2nd millennium BC.

These early uses appear to have been largely ceremonial or ornamental. Meteoritic iron is very rare, and the metal was probably very expensive, perhaps more expensive than gold. The early Hittites are known to have bartered iron (meteoritic or smelted) for silver, at a rate of 40 times the iron's weight, with Assyria.

Meteoric iron was also fashioned into tools in the Arctic, about the year 1000, when the Thule people of Greenland began making harpoons, knives, ulos and other edged tools from pieces of the Cape York meteorite. Typically pea-size bits of metal were cold-hammered into disks and fitted to a bone handle. These artifacts were also used as trade goods with other Arctic peoples: tools made from the Cape York meteorite have been found in archaeological sites more than 1,000 miles

(1,600 km) distant. When the American polar explorer Robert Peary shipped the largest piece of the meteorite to the American Museum of Natural History in New York City in 1897, it still weighed over 33 tons. Another example of a late use of meteoritic iron is an adze from around 1000 AD found in Sweden.

## Native Iron

Native iron in the metallic state occurs rarely as small inclusions in certain basalt rocks. Besides meteoritic iron, Thule people of Greenland have used native iron from the Disko region.

## Iron Smelting and the Iron Age

Iron smelting—the extraction of usable metal from oxidized iron ores—is more difficult than tin and copper smelting. While these metals and their alloys can be cold-worked or melted in relatively simple furnaces (such as the kilns used for pottery) and cast into molds, smelted iron requires hot-working and can be melted only in specially designed furnaces. Thus it is not surprising that humans mastered the technology of smelted iron only after several millennia of bronze metallurgy.

The place and time for the discovery of iron smelting is not known, partly because of the difficulty of distinguishing metal extracted from nickel-containing ores from hot-worked meteoritic iron. The archaeological evidence seems to point to the Middle East area, during the Bronze Age in the 3rd millennium BC. However, iron artifacts remained a rarity until the 12th century BC.

The Iron Age is conventionally defined by the widespread replacement of bronze weapons and tools with those of steel. That transition happened at different times in different places, as the technology spread. Mesopotamia was fully into the Iron Age by 900 BC. Although Egypt produced iron artifacts, bronze remained dominant until its conquest by Assyria in 663 BC. The Iron Age began in Central Europe about 500 BC, and in India and China between 1200 and 500 BC. Around 500 BC, the Nubians who had learned from the Assyrians the use of iron and were expelled from Egypt, became major manufacturers and exporters of iron.

## Ancient Near East

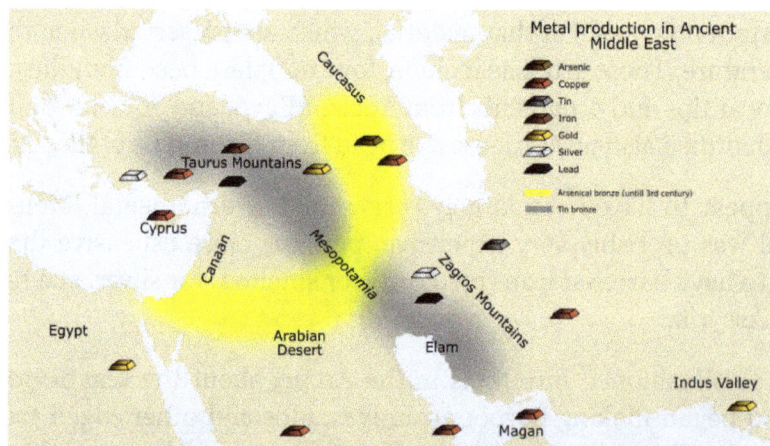

Mining areas of the ancient Middle East. Boxes colors: arsenic is in brown, copper in red, tin in grey, iron in reddish brown, gold in yellow, silver in white and lead in black. Yellow area stands for arsenic bronze, while grey area stands for tin bronze

One of the earliest smelted iron artifacts, a dagger with an iron blade found in a Hattic tomb in Anatolia, dated from 2500 BC. About 1500 BC, increasing numbers of non-meteoritic, smelted iron objects appeared in Mesopotamia, Anatolia, and Egypt. Nineteen iron objects were found in the tomb of Egyptian ruler Tutankhamun, died in 1323 BC, including an iron dagger with a golden hilt, an Eye of Horus, the mummy's head-stand and sixteen models of an artisan's tools. An Ancient Egyptian sword bearing the name of pharaoh Merneptah as well as a battle axe with an iron blade and gold-decorated bronze shaft were both found in the excavation of Ugarit.

Although iron objects dating from the Bronze Age have been found across the Eastern Mediterranean, bronzework appears to have greatly predominated during this period. By the 12th century BC, iron smelting and forging, of weapons and tools, was common from Sub-Saharan Africa through India. As the technology spread, iron came to replace bronze as the dominant metal used for tools and weapons across the Eastern Mediterranean (the Levant, Cyprus, Greece, Crete, Anatolia, and Egypt).

Iron was originally smelted in bloomeries, furnaces where bellows were used to force air through a pile of iron ore and burning charcoal. The carbon monoxide produced by the charcoal reduced the iron oxide from the ore to metallic iron. The bloomery, however, was not hot enough to melt the iron, so the metal collected in the bottom of the furnace as a spongy mass, or *bloom*. Workers then repeatedly beat and folded it to force out the molten slag. This laborious, time-consuming process produced wrought iron, a malleable but fairly soft alloy.

Concurrent with the transition from bronze to iron was the discovery of carburization, the process of adding carbon to wrought iron. While the iron bloom contained some carbon, the subsequent hot-working oxidized most of it. Smiths in the Middle East discovered that wrought iron could be turned into a much harder product by heating the finished piece in a bed of charcoal, and then quenching it in water or oil. This procedure turned the outer layers of the piece into steel, an alloy of iron and iron carbides, with an inner core of less brittle iron.

## Theories on the Origin of Iron Smelting

The development of iron smelting was traditionally attributed to the Hittites of Anatolia of the Late Bronze Age. It was believed that they maintained a monopoly on iron working, and that their empire had been based on that advantage. According to that theory, the ancient Sea Peoples, who invaded the Eastern Mediterranean and destroyed the Hittite empire at the end of the Late Bronze Age, were responsible for spreading the knowledge through that region. This theory is no longer held in the mainstream of scholarship, since there is no archaeological evidence of the alleged Hittite monopoly. While there are some iron objects from Bronze Age Anatolia, the number is comparable to iron objects found in Egypt and other places of the same time period, and only a small number of those objects were weapons.

A more recent theory claims that the development of iron technology was driven by the disruption of the copper and tin trade routes, due to the collapse of the empires at the end of the Late Bronze Age. These metals, especially tin, were not widely available and metal workers had to transport them over long distances, whereas iron ores were widely available. However, no known archaeological evidence suggests a shortage of bronze or tin in the Early Iron Age. Bronze objects remained abundant, and these objects have the same percentage of tin as those from the Late Bronze Age.

## Indian Sub-Continent

The Iron pillar of Delhi

The history of metallurgy in the Indian subcontinent began in the 2nd millennium BC. Archaeological sites in Gangetic plains have yielded iron implements dated between 1800 – 1200 BC. By the early 13th century BC, iron smelting was practiced on a large scale in India. In Southern India (present day Mysore) iron was in use 12th to 11th centuries BC. The technology of iron metallurgy advanced in the politically stable Maurya period. and during a period of peaceful settlements in the 1st millennium BC.

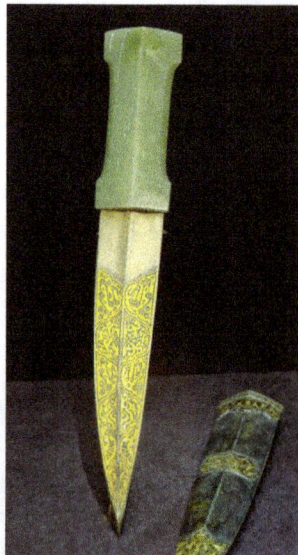

Dagger and its scabbard, India, 17th–18th century. Blade: Damascus steel inlaid with gold; hilt: jade; scabbard: steel with engraved, chased and gilded decoration

Iron artifacts such as spikes, knives, daggers, arrow-heads, bowls, spoons, saucepans, axes, chisels, tongs, door fittings etc., dated from 600 to 200 BC, have been discovered at several archaeological sites of India. The Greek historian Herodotus wrote the first western account of the use of iron in India. The Indian mythological texts, the Upanishads, have mentions of weaving, pottery, and metallurgy as well. The Romans had high regard for the excellence of steel from India in the time of the Gupta Empire.

Perhaps as early as 500 BC, although certainly by 200 AD, high quality steel was produced in southern India by the crucible technique. In this system, high-purity wrought iron, charcoal, and glass were mixed in a crucible and heated until the iron melted and absorbed the carbon. Iron chain was used in Indian suspension bridges as early as the 4th century.

Wootz steel was produced in India and Sri Lanka from around 300 BC. Wootz steel is famous from Classical Antiquity for its durability and ability to hold an edge. When asked by King Porus to select a gift, Alexander is said to have chosen, over gold or silver, thirty pounds of steel. Wootz steel was originally a complex alloy with iron as its main component together with various trace elements. Recent studies have suggested that its qualities may have been due to the formation of carbon nanotubes in the metal. According to Will Durant, the technology passed to the Persians and from them to Arabs who spread it through the Middle East. In the 16th century, the Dutch carried the technology from South India to Europe, where it was mass-produced.

Steel was produced in Sri Lanka from 300 BC by furnaces blown by the monsoon winds.The furnaces were dug into the crests of hills, and the wind was diverted into the air vents by long trenches. This arrangement created a zone of high pressure at the entrance, and a zone of low pressure at the top of the furnace. The flow is believed to have allowed higher temperatures than bellows-driven furnaces could produce, resulting in better-quality iron.Steel made in Sri Lanka was traded extensively within the region and in the Islamic world.

One of the world's foremost metallurgical curiosities is an iron pillar located in the Qutb complex, Delhi. The pillar is made of wrought iron (98% Fe), is almost seven meters high and weighs more than six tonnes. The pillar was erected by Chandragupta II Vikramaditya and has withstood 1,600 years of exposure to heavy rains with relatively little corrosion.

## Iron Age Europe

Axe made of iron, dating from Swedish Iron Age, found at Gotland, Sweden

Iron working was introduced to Greece in the late 10th century BC. The earliest marks of Iron Age in Central Europe are artifacts from the Hallstatt C culture (8th century BC). Throughout the 7th to 6th centuries BC, iron artifacts remained luxury items reserved for an elite. This changed dramatically shortly after 500 BC with the rise of the La Tène culture, from which time iron metallurgy also became common in Northern Europe and Britain. The spread of ironworking in Central and Western Europe is associated with Celtic expansion. By the 1st century BC, Noric steel was famous for its quality and sought-after by the Roman military.

The annual iron output of the Roman Empire is estimated at 84,750 t, while the similarly populous Han China produced around 5,000 t.

## China

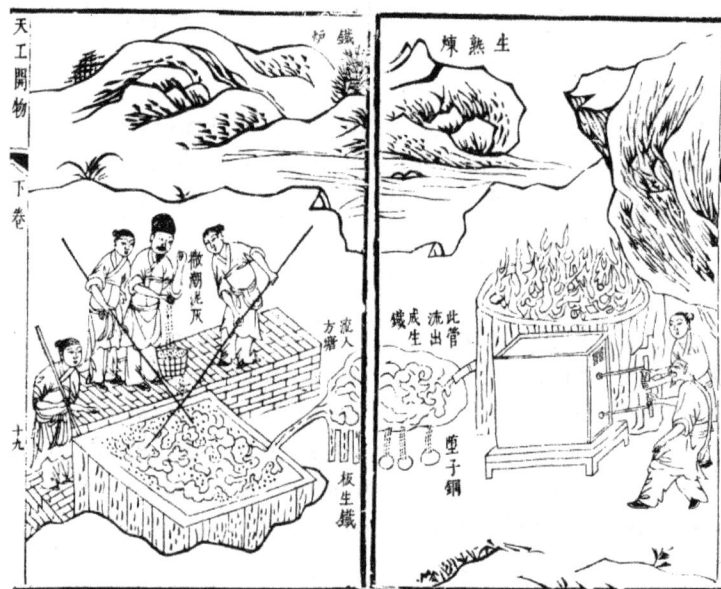

The process of fining iron ore to make wrought iron from pig iron, with the right illustration displaying men working a blast furnace, from the *Tiangong Kaiwu* encyclopedia, 1637

Historians debate whether bloomery-based ironworking ever spread to China from the Middle East. One theory suggests that metallurgy was introduced through Central Asia. The earliest cast iron artifacts, dating to 5th century BC, were discovered by archaeologists in what is now modern Luhe County, Jiangsu in China. Cast iron was used in ancient China for warfare, agriculture, and architecture. Around 500 BC, metalworkers in the southern state of Wu achieved a temperature of 1130 °C. At this temperature, iron combines with 4.3% carbon and melts. The liquid iron can be cast into molds, a method far less laborious than individually forging each piece of iron from a bloom. This technology would be known in Europe from early medieval times on.

Cast iron is rather brittle and unsuitable for striking implements. It can, however, be *decarburized* to steel or wrought iron by heating it in air for several days. In China, these iron working methods spread northward, and by 300 BC, iron was the material of choice throughout China for most tools and weapons. A mass grave in Hebei province, dated to the early 3rd century BC, contains several soldiers buried with their weapons and other equipment. The artifacts recovered from this grave are variously made of wrought iron, cast iron, malleabilized cast iron, and quench-hardened steel, with only a few, probably ornamental, bronze weapons.

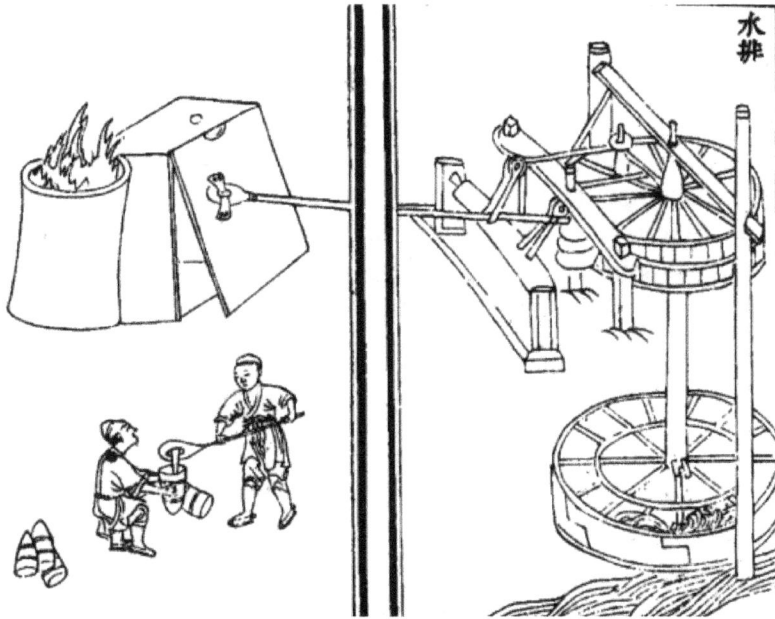

An illustration of furnace bellows operated by waterwheels, from the *Nong Shu*, by Wang Zhen, 1313 AD, during the Yuan Dynasty in China

During the Han Dynasty (202 BC–220 AD), the government established ironworking as a state monopoly (repealed during the latter half of the dynasty and returned to private entrepreneurship) and built a series of large blast furnaces in Henan province, each capable of producing several tons of iron per day. By this time, Chinese metallurgists had discovered how to fine molten pig iron, stirring it in the open air until it lost its carbon and could be hammered (wrought). (In modern Mandarin-Chinese, this process is now called *chao*, literally, stir frying.) By the 1st century BC, Chinese metallurgists had found that wrought iron and cast iron could be melted together to yield an alloy of intermediate carbon content, that is, steel. According to legend, the sword of Liu Bang, the first Han emperor, was made in this fashion. Some texts of the era mention "harmonizing the hard and the soft" in the context of ironworking; the phrase may refer to this process. The ancient city of Wan (Nanyang) from the Han period forward was a major center of the iron and steel industry. Along with their original methods of forging steel, the Chinese had also adopted the production methods of creating Wootz steel, an idea imported from India to China by the 5th century AD. During Han Dynasty, the Chinese were also the first to apply hydraulic power (i.e. a waterwheel) in working the bellows of the blast furnace. This was recorded in the year 31 AD, as an innovation of the engineer Du Shi, Prefect of Nanyang. Although Du Shi was the first to apply water power to bellows in metallurgy, the first drawn and printed illustration of its operation with water power appeared in 1313 AD, in the Yuan Dynasty era text called the *Nong Shu*. In the 11th century, there is evidence of the production of steel in Song China using two techniques: a "berganesque" method that produced inferior, heterogeneous steel and a precursor to the modern Bessemer process that utilized partial decarbonization via repeated forging under a cold blast. By the 11th century, there was a large amount of deforestation in China due to the iron industry's demands for charcoal. By this time however, the Chinese had learned to use bituminous coke to replace charcoal, and with this switch in resources many acres of prime timberland in China were spared. The change of fuel resources from charcoal to coal was pioneered in Roman Britain by the 2nd century AD, although it was also practiced in the Germanic Rhineland at the time.

## Africa South of the Sahara

Inhabitants at Termit, in eastern Niger became the first iron smelting people in West Africa around 1500 BC. Iron and copper working spread southward through the continent, reaching the Cape around AD 200. The widespread use of iron revolutionized the Bantu-speaking farming communities who adopted it, driving out and absorbing the rock tool using hunter-gatherer societies they encountered as they expanded to farm wider areas of savanna. The technologically superior Bantu-speakers spread across southern Africa and became wealthy and powerful, producing iron for tools and weapons in large, industrial quantities.

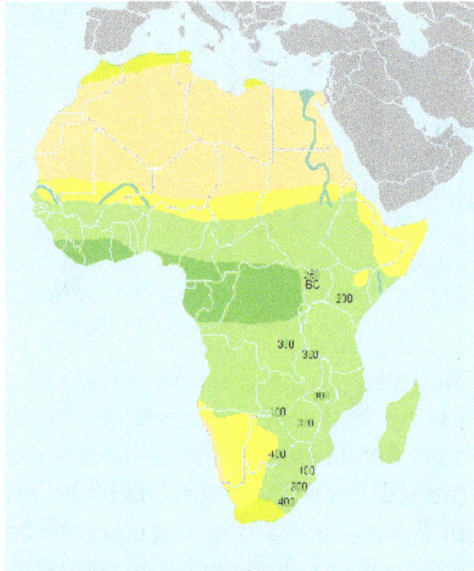

Iron Age finds in East and Southern Africa, corresponding to the early 1st millennium AD Bantu expansion

In the region of the Aïr Mountains in Niger there are signs of independent copper smelting between 2500–1500 BC. The process was not in a developed state, indicating smelting was not foreign. It became mature about the 1500 BC.

Similarly, smelting in bloomery-type furnaces in West Africa and forging for tools appear in the Nok culture in Africa by 500 BC. The earliest records of bloomery-type furnaces in East Africa are discoveries of smelted iron and carbon in Nubia and Axum that date back between 1,000-500 BCE. Particularly in Meroe, there are known to have been ancient bloomeries that produced metal tools for the Nubians and Kushites and produced surplus for their economy.

In the regions of Tanzania inhabited by the Haya people, carbon dating has shown that blast furnaces were as old as 2000 years, whereas steel of this calibre did not appear in Europe until several centuries later.

## Medieval Islamic World

Iron technology was further advanced by several inventions in medieval Islam, during the so-called Islamic Golden Age. These included a variety of water-powered and wind-powered industrial mills for metal production, including geared gristmills and forges. By the 11th century, every province throughout the Muslim world had these industrial mills in operation, from Islamic

Spain and North Africa in the west to the Middle East and Central Asia in the east. There are also 10th-century references to cast iron, as well as archeological evidence of blast furnaces being used in the Ayyubid and Mamluk empires from the 11th century, thus suggesting a diffusion of Chinese metal technology to the Islamic world.

Geared gristmills were invented by Muslim engineers, and were used for crushing metallic ores before extraction. Gristmills in the Islamic world were often made from both watermills and windmills. In order to adapt water wheels for gristmilling purposes, cams were used for raising and releasing trip hammers. The first forge driven by a hydropowered water mill rather than manual labour was invented in the 12th century Islamic Spain.

One of the most famous steels produced in the medieval Near East was Damascus steel used for swordmaking, and mostly produced in Damascus, Syria, in the period from 900 to 1750. This was produced using the crucible steel method, based on the earlier Indian wootz steel. This process was adopted in the Middle East using locally produced steels. The exact process remains unknown, but it allowed carbides to precipitate out as micro particles arranged in sheets or bands within the body of a blade. Carbides are far harder than the surrounding low carbon steel, so swordsmiths could produce an edge that cut hard materials with the precipitated carbides, while the bands of softer steel let the sword as a whole remain tough and flexible. A team of researchers based at the Technical University of Dresden that uses X-rays and electron microscopy to examine Damascus steel discovered the presence of cementite nanowires and carbon nanotubes. Peter Paufler, a member of the Dresden team, says that these nanostructures give Damascus steel its distinctive properties and are a result of the forging process.

## Medieval and Early Modern Europe

There was no fundamental change in the technology of iron production in Europe for many centuries. European metal workers continued to produce iron in bloomeries. However, the Medieval period brought two developments—the use of water power in the bloomery process in various places (outlined above), and the first European production in cast iron.

## Powered Bloomeries

Sometime in the medieval period, water power was applied to the bloomery process. It is possible that this was at the Cistercian Abbey of Clairvaux as early as 1135, but it was certainly in use in early 13th century France and Sweden. In England, the first clear documentary evidence for this is the accounts of a forge of the Bishop of Durham, near Bedburn in 1408, but that was certainly not the first such ironworks. In the Furness district of England, powered bloomeries were in use into the beginning of the 18th century, and near Garstang until about 1770.

The Catalan Forge was a variety of powered bloomery. Bloomeries with hot blast were used in up-state New York in the mid-19th century.

## Blast Furnace

Cast iron development lagged in Europe, as the smelters could only achieve temperatures of about 1000 C; or perhaps they did not want hotter temperatures, as they were seeking to produce blooms as a precursor of wrought iron, not cast iron. Through a good portion of the Middle Ages, in Western Europe, iron was still being made by the working of iron blooms into wrought iron. Some of the earliest casting of iron in Europe occurred in Sweden, in two sites, Lapphyttan and Vinarhyttan, between 1150 and 1350. Some scholars have speculated the practice followed the Mongols across Russia to these sites, but there is no clear proof of this hypothesis, and it would certainly not explain the pre-Mongol datings of many of these iron-production centres. In any event, by the late 14th century, a market for cast iron goods began to form, as a demand developed for cast iron cannonballs.

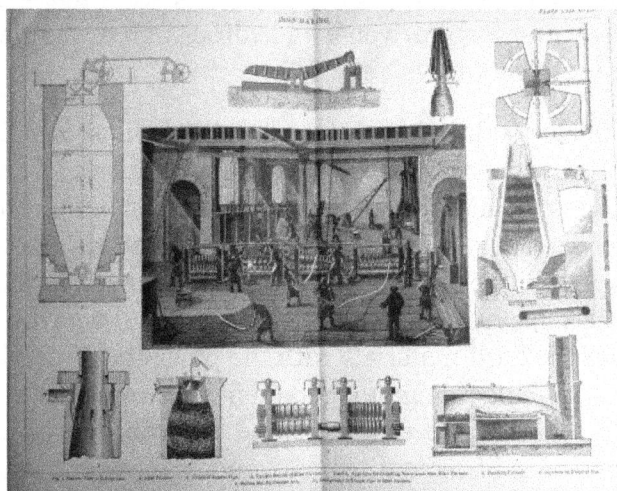

Ironmaking described in "The Popular Encyclopedia" vol.VII, published 1894

## Osmond Process

Iron from furnaces such as Lapphyttan was refined into wrought iron by the osmond process. The pig iron from the furnace was melted in front of a blast of air and the droplets caught on a staff (which was spun). This formed a ball of iron, known as an osmond. This was probably a traded commodity by c. 1200.

## Finery Process

An alternative method of decarburising pig iron was the finery process, which seems to have been devised in the region around Namur in the 15th century. By the end of that century, this Walloon process spread to the *Pay de Bray* on the eastern boundary of Normandy, and then to England, where it became the main method of making wrought iron by 1600. It was introduced to Sweden by Louis de Geer in the early 17th century and was used to make the oregrounds iron favoured by English steelmakers.

A variation on this was the German process. This became the main method of producing bar iron in Sweden.

## Cementation Steel

In the early 17th century, ironworkers in Western Europe had developed the cementation process for carburizing wrought iron. Wrought iron bars and charcoal were packed into stone boxes, then held at a red heat for up to a week. During this time, carbon diffused into the iron, producing a product called *cement steel* or *blister steel*. One of the earliest places where this was used in England was at Coalbrookdale, where Sir Basil Brooke had two cementation furnaces (recently excavated). For a time in the 1610s, he owned a patent on the process, but had to surrender this in 1619. He probably used Forest of Dean iron as his raw material, but it was soon found that oregrounds iron was more suitable. The quality of the steel could be improved by faggoting, producing the so-called shear steel.

## Crucible Steel

In the 1740s, Benjamin Huntsman found a means of melting blister steel, made by the cementation process, in crucibles. The resulting crucible steel, usually cast in ingots, was more homogeneous than blister steel.

## Transition to Coke in England

### Beginnings

Early iron smelting used charcoal as both the heat source and the reducing agent. By the 18th century, the availability of wood for making charcoal was limiting the expansion of iron production, so that England became increasingly dependent for a considerable part of the iron required by its industry, on Sweden (from the mid-17th century) and then from about 1725 also on Russia.

Smelting with coal (or its derivative coke) was a long sought objective. The production of pig iron with coke was probably achieved by Dud Dudley in the 1620s, and with a mixed fuel made from coal and wood again in the 1670s. However this was probably only a technological rather than a commercial success. Shadrach Fox may have smelted iron with coke at Coalbrookdale in Shropshire in the 1690s, but only to make cannonballs and other cast iron products such as shells. However, in the peace after the Nine Years War, there was no demand for these.

### Abraham Darby and his Successors

In 1707, Abraham Darby patented a method of making cast iron pots. His pots were thinner and hence cheaper than those of his rivals. Needing a larger supply of pig iron he leased the blast furnace at Coalbrookdale in 1709. There, he made iron using coke, thus establishing the first successful business in Europe to do so. His products were all of cast iron, though his immediate successors attempted (with little commercial success) to fine this to bar iron.

Bar iron thus continued normally to be made with charcoal pig iron until the mid-1750s. In 1755 Abraham Darby II (with partners) opened a new coke-using furnace at Horsehay in Shropshire, and this was followed by others. These supplied coke pig iron to finery forges of the traditional kind for the production of bar iron. The reason for the delay remains controversial.

### New Forge Processes

It was only after this that economically viable means of converting pig iron to bar iron began to be devised. A process known as potting and stamping was devised in the 1760s and improved in the 1770s, and seems to have been widely adopted in the West Midlands from about 1785. However, this was largely replaced by Henry Cort's puddling process, patented in 1784, but probably only made to work with grey pig iron in about 1790. These processes permitted the great expansion in the production of iron that constitutes the Industrial Revolution for the iron industry.

Schematic drawing of a puddling furnace

In the early 19th century, Hall discovered that the addition of iron oxide to the charge of the puddling furnace caused a violent reaction, in which the pig iron was decarburised, this became known as 'wet puddling'. It was also found possible to produce steel by stopping the puddling process before decarburisation was complete.

## Hot Blast

The efficiency of the blast furnace was improved by the change to hot blast, patented by James Beaumont Neilson in Scotland in 1828. This further reduced production costs. Within a few decades, the practice was to have a 'stove' as large as the furnace next to it into which the waste gas (containing CO) from the furnace was directed and burnt. The resultant heat was used to preheat the air blown into the furnace.

## Industrial Steelmaking

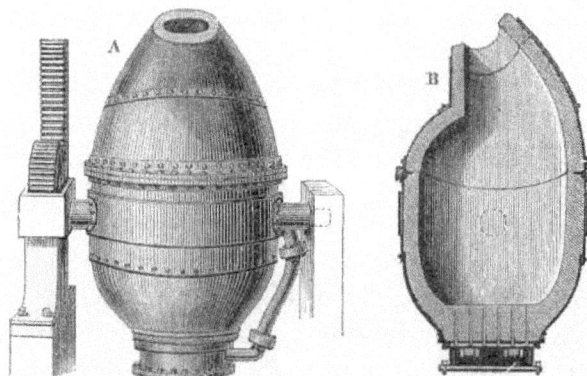

Schematic drawing of a Bessemer converter

Apart from some production of puddled steel, English steel continued to be made by the cementation process, sometimes followed by remelting to produce crucible steel. These were batch-based processes whose raw material was bar iron, particularly Swedish oregrounds iron.

The problem of mass-producing cheap steel was solved in 1855 by Henry Bessemer, with the introduction of the Bessemer converter at his steelworks in Sheffield, England. (An early converter can still be seen at the city's Kelham Island Museum). In the Bessemer process, molten pig iron from the blast furnace was charged into a large crucible, and then air was blown through the molten iron from below, igniting the dissolved carbon from the coke. As the carbon burned off, the melting point of the mixture increased, but the heat from the burning carbon provided the extra energy needed to keep the mixture molten. After the carbon content in the melt had dropped to the desired level, the air draft was cut off: a typical Bessemer converter could convert a 25-ton batch of pig iron to steel in half an hour.

Finally, the basic oxygen process was introduced at the Voest-Alpine works in 1952; a modification of the basic Bessemer process, it lances oxygen from above the steel (instead of bubbling air from below), reducing the amount of nitrogen uptake into the steel. The basic oxygen process is used in all modern steelworks; the last Bessemer converter in the U.S. was retired in 1968. Furthermore, the last three decades have seen a massive increase in the mini-mill business, where scrap steel only is melted with an electric arc furnace. These mills only produced bar products at first, but have since expanded into flat and heavy products, once the exclusive domain of the integrated steelworks.

Until these 19th-century developments, steel was an expensive commodity and only used for a limited number of purposes where a particularly hard or flexible metal was needed, as in the cutting edges of tools and springs. The widespread availability of inexpensive steel powered the Second Industrial Revolution and modern society as we know it. Mild steel ultimately replaced wrought iron for almost all purposes, and wrought iron is no longer commercially produced. With minor exceptions, alloy steels only began to be made in the late 19th century. Stainless steel was developed on the eve of the First World War and was not widely used until the 1920s.

## Non-ferrous Metal

In metallurgy, a non-ferrous metal is a metal, including alloys, that does not contain iron (ferrite) in appreciable amounts. Generally more expensive than ferrous metals, non-ferrous metals are used because of desirable properties such as low weight (e.g. aluminium), higher conductivity (e.g. copper), non-magnetic property or resistance to corrosion (e.g. zinc). Some non-ferrous materials are also used in the iron and steel industries. For example, bauxite is used as flux for blast furnaces, while others such as wolframite, pyrolusite and chromite are used in making ferrous alloys.

Important non-ferrous metals include aluminium, copper, lead, nickel, tin, titanium and zinc, and alloys such as brass. Precious metals such as gold, silver and platinum and exotic or rare metals such as cobalt, mercury, tungsten, beryllium, bismuth, cerium, cadmium, niobium, indium, gallium, germanium, lithium, selenium, tantalum, tellurium, vanadium, and zirconium are also non-ferrous. They are usually obtained through minerals such as sulfides, carbonates, and silicates. Non-ferrous metals are usually refined through electrolysis.

## Recycling and Pollution Control

Due to their extensive use, non-ferrous scrap metals are usually recycled. The secondary materials in scrap are vital to the metallurgy industry, as the production of new metals often needs them. Some recycling facilities re-smelt and recast non-ferrous materials; the dross is collected and stored onsite while the metal fumes are filtered and collected. Non-ferrous scrap metals are sourced from industrial scrap materials, particle emissions and obsolete technology (for example, copper cables) scrap.

## Ancient History

Non-ferrous metals were the first metals used by humans for metallurgy. Gold, silver and copper existed in their native crystalline yet metallic form. These crystals, though rare, are enough to attract the attention of humans. Less susceptible to oxygen than most other metals, they can be found even in weathered outcroppings. Copper was the first metal to be forged; it was soft enough to be fashioned into various objects by cold forging, and it could be melted in a crucible. Gold, silver and copper replaced some of the functions of other resources, such as wood and stone, owing to their ability to be shaped into various forms for different uses. Due to their rarity, these gold, silver and copper artifacts were treated as luxury items and handled with great care. The use of copper also heralded the transition from the Stone Age to the Copper Age. The Bronze Age, which succeeded the Copper Age, was again heralded by the invention of bronze, an alloy of copper with the non-ferrous metal tin.

## Mechanical and Structural Use

It is used in residential, commercial, industrial industry. Material selection for a mechanical or structural application requires some important considerations, including how easily the material can be shaped into a finished part and how its properties can be either intentionally or inadvertently altered in the process. Depending on the end use, metals can be simply cast into the finished part, or cast into an intermediate form, such as an ingot, then worked, or wrought, by rolling, forging, extruding, or other deformation process. Although the same operations are used with ferrous as well as nonferrous metals and alloys, the reaction of nonferrous metals to these forming processes is often more severe. Consequently, properties may differ considerably between the cast and wrought forms of the same metal or alloy.

## References

- Davis, Joseph R. (1993) ASM Specialty Handbook: Aluminum and Aluminum Alloys. ASM International. p. 211. ISBN 978-0-87170-496-2.

- Miskimin, Harry A. (1977) The economy of later Renaissance Europe, 1460–1600. Cambridge University Press. p. 31. ISBN 0-521-29208-5.

- Nicholson, Paul T. and Shaw, Ian (2000) Ancient Egyptian materials and technology. Cambridge University Press. pp. 164–167. ISBN 0-521-45257-0.

- Roberts, George Adam; Krauss, George; Kennedy, Richard and Kennedy, Richard L. (1998) Tool steels. ASM International. pp. 2–3. ISBN 0-87170-599-0.

- Bramfitt, B. L. (2001). Metallographer's Guide: Practice and Procedures for Irons and Steels. ASM International. pp. 13–. ISBN 978-1-61503-146-7.

- Gupta, R. C. (2010). Theory and laboratory experiments in ferrous metallurgy. New Delhi: PHI Learning Private Ltd. p. 6. ISBN 978-81-203-3924-8.

- Young, Courtney A., ed. (2008). Hydrometallurgy 2008 : proceedings of the sixth international symposium (1st ed.). Littleton, Colo.: Society for Mining, Metallurgy, and Exploration. p. 416. ISBN 978-0-87335-266-6.

- "Commonly Recycled Metals and Their Sources" (PDF). Occupational Safety and Health Administration. Retrieved 27 October 2011.

- "Chapter 82 – Metal Processing and Metal Working Industry". Encyclopaedia of Occupational Health and Safety, 4th Edition. Retrieved 26 October 2011.

- "Department of the Environment Industry Profile: Waste recycling, treatment and disposal sites" (PDF). Environment Agency. Retrieved 27 October 2011.

# 5

# An Overview of Ironworks

Due to its tremendous potential of use, iron is utilized in a wide variety of applications. This chapter details the various processes involved in the extraction of iron, namely smelting, etc. The content also has a section about the various furnaces used in the extraction of iron and ironworks. The chapter on ironworks offers an insightful focus, keeping in mind the complex subject matter.

## Ironworks

An ironworks or iron works is a building or site where iron is smelted and where heavy iron and steel products are made. The term is both singular and plural, i.e. the singular of *ironworks* is *ironworks*.

*The Iron Rolling Mill* (*Eisenwalzwerk*), 1870s, by Adolph Menzel.

Ironworks succeed bloomeries when blast furnaces replaced former methods. An integrated ironworks in the 19th century usually included one or more blast furnaces and a number of puddling furnaces or a foundry with or without other kinds of ironworks. After the invention of the Bessemer process, converters became widespread, and the appellation steelworks replaced ironworks.

The processes carried at ironworks are usually described as ferrous metallurgy, but the term **siderurgy** is also occasionally used. This is derived from the Greek words *sideros* - iron and *ergon* or *ergos* - work. This is an unusual term in English, and it is best regarded as an anglicisation of a term used in French, Spanish, and other Romance languages.

Casting at an iron foundry: *From Fra Burmeister og Wain's Iron Foundry*, 1885 by Peder Severin Krøyer

## Varieties of Ironworks

## Primary Ironmaking

A South Wales iron mill in 1798

Blast furnaces of Třinec Iron and Steel Works.

Toronto rolling mills

Ironworks is used as an omnibus term covering works undertaking one or more iron-producing processes. Such processes or species of ironworks where they were undertaken include the following:

- Blast furnaces — which made pig iron (or sometimes finished cast iron goods) from iron ore;

- Bloomeries — where bar iron was produced from iron ore by direct reduction;

- Electrolytic smelting — Employs a chromium/iron anode that can survive a 2,850 °F (1,570 °C) to produce decarbonized iron and 2/3 of a ton of industrial-quality oxygen per ton of iron. A thin film of metal oxide forms on the anode in the intense heat. The oxide forms a protective layer that prevents excess consumption of the base metal.

- Finery forges — which fined pig iron to produce bar iron, using charcoal as fuel in a finery (hearth) and coal or charcoal in a chafery (hearth);

- Foundries — where pig iron was remelted in an air furnace or in a foundry cupola to produce cast iron goods;

- Potting and stamping forges with melting fineries using the first process in which bar iron was made from pig iron with mineral coal or coke, without the use of charcoal;

- Puddling furnaces — a later process for the same purpose, again with coke as fuel. It was usually necessary for there to be a preliminary refining process in a coke refinery (also called running out furnace). After puddling, the puddled ball needed shingling (metallurgy) and then to be drawn out into bar iron in a rolling mills.

## Modern Steelmaking

From the 1850s, pig iron might be partly decarburised to produce mild steel using one of the following:

- The Bessemer process in a Bessemer converter, improved by the Gilchrist-Thomas process;

- The Siemens-Martin process in an Open hearth furnace;

- Electric arc furnace, introduced in 1907;

- Basic oxygen steelmaking, introduced in 1952.

The mills operating converters of any type are better called steelworks, ironworks referring to former processes, like puddling.

## Further Processing

After bar iron had been produced in a finery forge or in the forge train of a rolling mill, it might undergo further processes in one of the following:

- A slitting mill - which cut a flat bar into rod iron suitable for making into nails.

- A tinplate works - where rolling mills made sheets of iron (later of steel), which were coated with tin.

- A plating forge with a tilt hammer, a lighter hammer with a rapid stroke rate, enabling the production of thinner iron, suitable for the manufacture of knives, other cutlery, and so on.

- A cementation furnace might be used to convert the bar iron (if it was pure enough) into blister steel by the cementation process, either as an end in itself or as the raw material for crucible steel.

## Manufacture

Most of these processes did not produce finished goods. Further processes were often manual, including

- Manufacturing by blacksmiths or more specialist kind of smith.

- It might be used in shipbuilding.

In the context of the iron industry, the term *manufacture* is best reserved for this final stage.

## Notable Ironworks

Coat of arms of Eisenhüttenstadt ("city of ironworks"), Germany

## Great Britain

- Blaenavon Ironworks Heritage Site. Blaenavon (Blaenafon) South Wales

- Coalbrookdale Ironworks

- Cyfarthfa Ironworks at Merthyr Tydfil, Glamorgan, south Wales

- Dowlais Ironworks also at Merthyr Tydfil

- Millwall Ironworks, a shipbuilding firm on the Isle of Dogs, on the River Thames, London, England

- Thames Ironworks and Shipbuilding Co. Ltd, a shipbuilding firm at Leamouth on the River Thames, England

- Vulcan Iron Works at Bradford and other places

## United States of America

- Bath Iron Works in Maine

- Burden Iron Works in Troy, New York

- Cambria Iron Company in Johnstown, Pennsylvania

- Falling Creek Ironworks, Virginia.

- Saugus Iron Works in Saugus, Massachusetts

- Toledo Iron Works in Miami, Florida

- Tredegar Iron Works at Richmond, Virginia

- Vulcan Iron Works in Pennsylvania and other places

## Czech Republic

- Třinec Iron and Steel Works in Třinec, Czech Republic

## Germany

- Völklingen Ironworks Heritage Site

## Spain

- Altos Hornos de Vizcaya in Bilbao

- Arcelor facilities in Avilés and Gijón, formerly Ensidesa

## Historical

- Kindiba, in Burkina Faso. Ancient iron extraction site consisting of mines and three clay built furnaces.

- Darkhill Ironworks, in the Forest of Dean, England. Experimental ironworks established in 1818 and designated an 'Industrial Archaeological Site of International Importance'

- Royal Ironworks of St John, Ipanema, in São Paulo state, Brazil

# Smelting

Electric phosphate smelting furnace in a TVA chemical plant (1942)

Smelting is a form of extractive metallurgy; its main use is to produce a base metal from its ore. This includes production of silver, iron, copper and other base metals from their ores. Smelting makes use of heat and a chemical reducing agent to decompose the ore, driving off other elements as gases or slag and leaving just the metal base behind. The reducing agent is commonly a source of carbon such as coke, or in earlier times charcoal. The carbon (or carbon monoxide derived from it) removes oxygen from the ore, leaving behind the elemental metal. The carbon is thus oxidized in two stages, producing first carbon monoxide and then carbon dioxide. As most ores are impure, it is often necessary to use flux, such as limestone, to remove the accompanying rock gangue as slag.

Plants for the electrolytic reduction of aluminium are also generally referred to as aluminium smelters.

## Process

Smelting involves more than just melting the metal out of its ore. Most ores are a chemical compound of the metal with other elements, such as oxygen (as an oxide), sulfur (as a sulfide) or carbon and oxygen together (as a carbonate). To produce the metal, these compounds have to undergo a chemical reaction. Smelting therefore consists of using suitable reducing substances that will combine with those oxidizing elements to free the metal.

## Roasting

In the case of carbonates and sulfides, a process called "roasting" drives out the unwanted carbon or sulfur, leaving an oxide, which can be directly reduced. Roasting is usually carried out in an oxidizing environment. A few practical examples:

- Malachite, a common ore of copper, is primarily copper carbonate hydroxide $Cu_2(CO_3)$

(OH)$_2$. This mineral undergoes thermal decomposition to 2CuO, CO$_2$, and H$_2$O in several stages between 250 °C and 350 °C. The carbon dioxide and water are expelled into the atmosphere, leaving copper(II) oxide which can be directly reduced to copper as described in the following section titled Reduction.

- Galena, the most common mineral of lead, is primarily lead sulfide (PbS). The sulfide is oxidized to a sulfite (PbSO$_3$) which thermally decomposes into lead oxide and sulfur dioxide gas. (PbO and SO$_2$) The sulfur dioxide is expelled (like the carbon dioxide in the previous example), and the lead oxide is reduced as below.

## Reduction

Reduction is the final, high-temperature step in smelting. It is here that the oxide becomes the elemental metal. A reducing environment (often provided by carbon monoxide, made by incomplete combustion, produced in an air-starved furnace) pulls the final oxygen atoms from the raw metal. The required temperature varies over a very large range, both in absolute terms and in terms of the melting point of the base metal. A few examples:

- iron oxide becomes metallic iron at roughly 1250 °C (2282 °F or 1523.15 K), almost 300 degrees *below* iron's melting point of 1538 °C (2800.4 °F or 1811.15 K)

- mercuric oxide becomes vaporous mercury near 550 °C (1022 °F or 823.15 K), almost 600 degrees *above* mercury's melting point of -38 °C (-36.4 °F or 235.15 K)

Flux and slag can provide a secondary service after the reduction step is complete: They provide a molten cover on the purified metal, preventing it from coming into contact with oxygen while it is still hot enough to oxidize readily.

## Fluxes

Fluxes are used in smelting for several purposes, chief among them catalyzing the desired reactions and chemically binding to unwanted impurities or reaction products. Calcium oxide, in the form of lime, was often used for this purpose, since it could react with the carbon dioxide and sulfur dioxide produced during roasting and smelting to keep them out of the working environment.

## History

Of the seven metals known in antiquity, only gold occurred regularly in native form in the natural environment. The others – copper, lead, silver, tin, iron and mercury – occur primarily as minerals, though copper is occasionally found in its native state in commercially significant quantities. These minerals are primarily carbonates, sulfides, or oxides of the metal, mixed with other components such as silica and alumina. Roasting the carbonate and sulfide minerals in air converts them to oxides. The oxides, in turn, are smelted into the metal. Carbon monoxide was (and is) the reducing agent of choice for smelting. It is easily produced during the heating process, and as a gas comes into intimate contact with the ore.

In the Old World, humans learned to smelt metals in prehistoric times, more than 8000 years ago. The discovery and use of the "useful" metals — copper and bronze at first, then iron a few millennia later — had an enormous impact on human society. The impact was so pervasive that scholars traditionally divide ancient history into Stone Age, Bronze Age, and Iron Age.

In the Americas, pre-Inca civilizations of the central Andes in Peru had mastered the smelting of copper and silver at least six centuries before the first Europeans arrived in the 16th century, while never mastering the smelting of metals such as iron for use with weapon-craft.

## Tin and Lead

In the Old World, the first metals smelted were tin and lead. The earliest known cast lead beads were found in the Çatal Höyük site in Anatolia (Turkey), and dated from about 6500 BC, but the metal may have been known earlier.

Since the discovery happened several millennia before the invention of writing, there is no written record about how it was made. However, tin and lead can be smelted by placing the ores in a wood fire, leaving the possibility that the discovery may have occurred by accident.

Although lead is a common metal, its discovery had relatively little impact in the ancient world. It is too soft to be used for structural elements or weapons, except for the fact that it is exceptionally dense, making it ideal for sling projectiles. However, being easy to cast and shape, it came to be extensively used in the classical world of Ancient Greece and Ancient Rome for piping and storage of water. It was also used as a mortar in stone buildings.

Tin was much less common than lead and is only marginally harder, and had even less impact by itself.

## Copper and Bronze

After tin and lead, the next metal to be smelted appears to have been copper. How the discovery came about is a matter of much debate. Campfires are about 200 °C short of the temperature needed for that, so it has been conjectured that the first smelting of copper may have been achieved in pottery kilns. The development of copper smelting in the Andes, which is believed to have occurred independently of that in the Old World, may have occurred in the same way. The earliest current evidence of copper smelting, dating from between 5500 BC and 5000 BC, has been found in Pločnik and Belovode, Serbia. A mace head found in Can Hasan, Turkey and dated to 5000 BC, once thought to be the oldest evidence, now appears to be hammered native copper.

By combining copper with tin and/or arsenic in the right proportions one obtains bronze, an alloy which is significantly harder than copper. The first copper/arsenic bronzes date from 4200 BC from Asia Minor. The Inca bronze alloys were also of this type. Arsenic is often an impurity in copper ores, so the discovery could have been made by accident; but eventually arsenic-bearing minerals were intentionally added during smelting.

Copper–tin bronzes, harder and more durable, were developed around 3200 BC, also in Asia Minor.

The process through which the smiths learned to produce copper/tin bronzes is once again a mystery. The first such bronzes were probably a lucky accident from tin contamination of copper ores, but by 2000 BC, we know that tin was being mined on purpose for the production of bronze. This is amazing, given that tin is a semi-rare metal, and even a rich cassiterite ore only has 5% tin. Also, it takes special skills (or special instruments) to find it and to locate the richer lodes. But, whatever steps were taken to learn about tin, these were fully understood by 2000 BC.

The discovery of copper and bronze manufacture had a significant impact on the history of the Old World. Metals were hard enough to make weapons that were heavier, stronger, and more resistant to impact-related damage than their wood, bone, or stone equivalents. For several millennia, bronze was the material of choice for weapons such as swords, daggers, battle axes, and spear and arrow points, as well as protective gear such as shields, helmets, greaves (metal shin guards), and other body armor. Bronze also supplanted stone, wood, and organic materials in all sorts of tools and household utensils, such as chisels, saws, adzes, nails, blade shears, knives, sewing needles and pins, jugs, cooking pots and cauldrons, mirrors, horse harnesses, and much more. Tin and copper also contributed to the establishment of trade networks spanning large areas of Europe and Asia, and had a major effect on the distribution of wealth among individuals and nations.

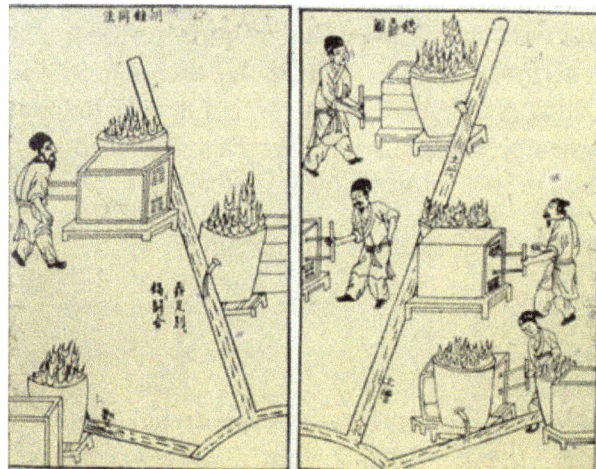

Casting bronze ding-tripods, from the Chinese *Tiangong Kaiwu* encyclopedia of Song Yingxing, published in 1637.

## Early Iron Smelting

Where and how iron smelting was discovered is widely debated, and remains uncertain due to the significant lack of production finds. Nevertheless, there is some consensus that iron technology originated in the Near East, perhaps in Eastern Anatolia.

In Ancient Egypt, somewhere between the Third Intermediate Period and 23rd Dynasty (ca. 1100–750 BC), there are indications of iron working. Significantly though, no evidence for the smelting of iron from ore has been attested to Egypt in any (pre-modern) period. There is a further possibility of iron smelting and working in West Africa by 1200 BC. In addition, very early instances of carbon steel were found to be in production around 2000 years before the present in northwest Tanzania, based on complex preheating principles. These discoveries are significant for the history of metallurgy.

Most early processes in Europe and Africa involved smelting iron ore in a bloomery, where the temperature is kept low enough so that the iron does not melt. This produces a spongy mass of iron called a bloom, which then has to be consolidated with a hammer. The earliest evidence to date for the bloomery smelting of iron is found at Tell Hammeh, Jordan, and dates to 930 BC (C14 dating).

## Later Iron Smelting

From the medieval period, the process of direct reduction in bloomeries began to be replaced by an indirect process. In this, a blast furnace was used to make pig iron, which then had to undergo a further process to make forgeable bar iron. Processes for the second stage include fining in a finery forge and, from the Industrial Revolution, puddling. However both processes are now obsolete, and wrought iron is now hardly made. Instead, mild steel is produced from a bessemer converter or by other means including smelting reduction processes such as the Corex Process.

## Base Metals

Cowles Syndicate of Ohio in Stoke-upon-Trent England, late 1880s. British Aluminium used the process of Paul Héroult about this time.

The ores of base metals are often sulfides. In recent centuries, reverberatory furnaces have been used. These keep the fuel and the charge being smelted separate. Traditionally these were used for carrying out the first step: formation of two liquids, one an oxide slag containing most of the impurity elements, and the other a sulfide matte containing the valuable metal sulfide and some impurities. Such "reverb" furnaces are today about 40 m long, 3 m high and 10 m wide. Fuel is burned at one end and the heat melts the dry sulfide concentrates (usually after partial roasting), which are fed through the openings in the roof of the furnace. The slag floats on top of the heavier matte, and is removed and discarded or recycled. The sulfide matte is then sent to the converter. The precise details of the process will vary from one furnace to another depending on the mineralogy of the orebody from which the concentrate originates.

While reverberatory furnaces were very good at producing slags containing very little copper, they were relatively energy inefficient and produced a low concentration of sulfur dioxide in their off-gases that made it difficult to capture, and consequently, they have been supplanted by a new generation of copper smelting technologies. More recent furnaces have been designed based upon bath smelting, top jetting lance smelting, flash smelting and blast furnaces. Some examples of bath smelters include the Noranda furnace, the Isasmelt furnace, the Teniente reactor, the Vunyukov smelter and the SKS technology to name a few. Top jetting lance smelters include the Mitsubishi smelting reactor. Flash smelters account for over 50% of the world's copper smelters. There are many more varieties of smelting processes, including the Kivset, Ausmelt, Tamano, EAF, and BF.

# Bloomery

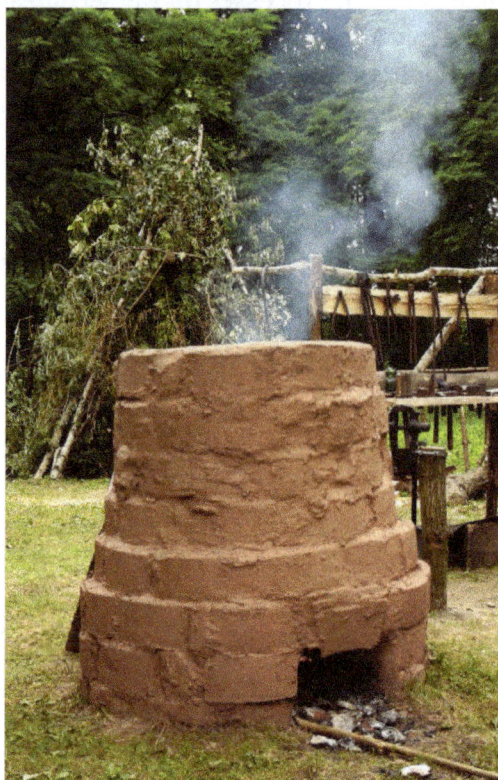

A bloomery in operation. The bloom will eventually be drawn out of the bottom hole.

A bloomery is a type of furnace once widely used for smelting iron from its oxides. The bloomery was the earliest form of smelter capable of smelting iron. A bloomery's product is a porous mass of iron and slag called a *bloom*. This mix of slag and iron in the bloom is termed *sponge iron*, which is usually consolidated and further forged into wrought iron. The bloomery has now largely been superseded by the blast furnace, which produces pig iron.

## Process

A bloomery consists of a pit or chimney with heat-resistant walls made of earth, clay, or stone. Near the bottom, one or more pipes (made of clay or metal) enter through the side walls. These pipes, called tuyères, allow air to enter the furnace, either by natural draught, or forced with bellows or a trompe. An opening at the bottom of the bloomery may be used to remove the bloom, or the bloomery can be tipped over and the bloom removed from the top.

The first step taken before the bloomery can be used is the preparation of the charcoal and the iron ore. The charcoal is produced by heating wood to produce the nearly pure carbon fuel needed for the smelting process. The ore is broken into small pieces and usually *roasted* in a fire to remove any moisture in the ore. Any large impurities in the ore can be crushed and removed. Since slag from previous blooms may have a high iron content, it can also be broken up and recycled into the bloomery with the new ore.

An iron bloom just removed from the furnace. Surrounding it are pieces of slag that have been pounded off by the hammer.

In operation, the bloomery is preheated by burning charcoal, and once hot, iron ore and additional charcoal are introduced through the top, in a roughly one to one ratio. Inside the furnace, carbon monoxide from the incomplete combustion of the charcoal reduces the iron oxides in the ore to metallic iron, without melting the ore; this allows the bloomery to operate at lower temperatures than the melting temperature of the ore. As the desired product of a bloomery is iron which is easily forgeable, it requires a low carbon content. The temperature and ratio of charcoal to iron ore must be carefully controlled to keep the iron from absorbing too much carbon and thus becoming unforgeable. Cast iron occurs when the iron melts and absorbs 2% to 4% carbon. Because the bloomery is self-fluxing the addition of limestone is not required to form a slag.

The small particles of iron produced in this way fall to the bottom of the furnace, where they combine with molten slag, often consisting of fayalite, a compound of silicon, oxygen and iron mixed with other impurities from the ore. The mixed iron and slag cool to form a spongy mass referred to as the bloom. Because the bloom is highly porous, and its open spaces are full of slag, the bloom must later be reheated and beaten with a hammer to drive the molten slag out of it. Iron treated this way is said to be *wrought* (worked), and the resulting iron, with reduced amounts of slag is called *wrought iron* or bar iron. It is also possible to produce blooms coated in steel by manipulating the charge of and air flow to the bloomery .

As the era of modern commercial steelmaking began, the word *bloom* was extended to another sense referring to an intermediate-stage piece of steel, of a size comparable to many traditional iron blooms, that was ready to be further worked into billet.

## History

A drawing of a simple bloomery and bellows.

Bloomery smelting during the Middle Ages.

The onset of the Iron Age in most parts of the world coincides with the first widespread use of the bloomery. While earlier examples of iron are found, their high nickel content indicates that this is meteoric iron. Other early samples of iron may have been produced by accidental introduction of iron ore in bronze smelting operations. Iron appears to have been smelted in the West as early as 3000 BC, but bronze smiths, not being familiar with iron, did not put it to use until much later. In the West, iron began to be used around 1200 BC.

## China

China has long been considered the exception to the general use of bloomeries. It was thought that the Chinese skipped the bloomery process completely, starting with the blast furnace and the finery forge to get wrought iron: by the 5th century BC, metalworkers in the southern state of Wu had invented the blast furnace, and the means to both cast iron and to decarburize the carbon-rich pig iron produced in a blast furnace to a low-carbon, wrought iron-like material. Recent evidence, however, shows that bloomeries were used earlier in China, migrating in from the west as early as 800 BC, before being supplanted by the locally developed blast furnace. Supporting this theory was the discovery of 'more than ten' iron digging elements found in the tomb of Duke Jing of Qin (d. 537 BCE), whose tomb is located in Fengxiang County, Shaanxi (a museum exists on the site today).

## Sub-Saharan Africa

Smelting in bloomery type furnaces in West Africa and forging of tools appeared in the Nok culture by 500 BC. The earliest records of bloomery-type furnaces in East Africa are discoveries of smelted iron and carbon in Nubia and Axum that dated to between 1,000–500 BC. In Meroe particularly there are known to have been ancient bloomeries that produced metal tools for the Nubians and Kushites and produced a surplus for sale.

## Medieval Europe

A Catalan forge, with tuyere and bellows on the right

Early European bloomeries were relatively small, smelting less than 1 kg of iron with each firing. Progressively larger bloomeries were constructed in the late fourteenth century, with a capacity of about 15 kg on average, though exceptions did exist. The use of waterwheels to power the bellows allowed the bloomery to become larger and hotter. European average bloom sizes quickly rose to 300 kg, where they levelled off until the demise of the bloomery.

As a bloomery's size is increased, the iron ore is exposed to burning charcoal for a longer time. When combined with the strong air blast required to penetrate the large ore and charcoal stack, this may cause part of the iron to melt and become saturated with carbon in the process, producing unforgeable pig iron which requires oxidation to be reduced into cast iron, steel, and iron. This pig iron was considered a waste product detracting from the largest bloomeries' yield, and it is not until the 14th century that early blast furnaces, identical in construction but dedicated to the production of molten iron, were built.

Bloomery type furnaces typically produced a range of iron products from very low carbon iron to steel containing approximately 0.2% to 1.5% carbon. The master smith had to select bits of low carbon iron, carburize them, and pattern-weld them together to make steel sheets. Even when applied to a non-carburized bloom, this pound, fold and weld process resulted in a more homogeneous product and removed much of the slag. The process had to be repeated up to 15 times when high quality steel was needed, as for a sword. The alternative was to carburize the surface of a finished product. Each welding's heat oxidises some carbon, so the master smith had to make sure there was enough carbon in the starting mixture.

In England and Wales, despite the arrival of the blast furnace in the Weald in about 1491, bloomery forges, probably using water-power for the hammer as well as the bellows, were operating in the West Midlands region beyond 1580. In Furness and Cumberland, they operated into the early 17th century and the last one in England (near Garstang) did not close until about 1770.

### Later Variations

Bloomeries survived in Spain and southern France as Catalan forges to the mid-19th century, and in Austria as the Stückofen to 1775.

A view of the bloomeries ( *'Catalan forges'* ) at Mission San Juan Capistrano, the oldest (*circa* 1790s) existing facilities of their kind in California.

In the Spanish colonization of the Americas, bloomeries or "Catalan forges" were part of 'self sufficiency' at some of the missions, encomiendas, and pueblos. As part of the Franciscan Spanish missions in Alta California, the "Catalan forges" at Mission San Juan Capistrano from the 1790s are the oldest existing facilities of their kind in the present day state of California. The bloomeries' sign proclaims the site as being "...part of Orange County's first industrial complex." In the Adirondacks, New York, new bloomeries using the hot blast technique were built in the 19th century.

## Blast Furnace

A blast furnace is a type of metallurgical furnace used for smelting to produce industrial metals, generally iron, but also others such as lead or copper.

Blast furnace in Sestao, Spain. The furnace itself is inside the central girderwork.

In a blast furnace, fuel, ores, and flux (limestone) are continuously supplied through the top of the furnace, while a hot blast of air (sometimes with oxygen enrichment) is blown into the lower section of the furnace through a series of pipes called tuyeres, so that the chemical reactions take place throughout the furnace as the material moves downward. The end products are usually molten metal and slag phases tapped from the bottom, and flue gases exiting from the top of the furnace. The downward flow of the ore and flux in contact with an upflow of hot, carbon monoxide-rich combustion gases is a countercurrent exchange process.

Part of the gas cleaning system of a blast furnace in Monclova, Mexico. This one is about to be de-commissioned and replaced.

In contrast, air furnaces (such as reverberatory furnaces) are naturally aspirated, usually by the convection of hot gases in a chimney flue. According to this broad definition, bloomeries for iron, blowing houses for tin, and smelt mills for lead would be classified as blast furnaces. However, the term has usually been limited to those used for smelting iron ore to produce pig iron, an intermediate material used in the production of commercial iron and steel, and the shaft furnaces used in combination with sinter plants in base metals smelting.

## History

Blast furnaces existed in China from about 1st century AD and in the West from the High Middle Ages. They spread from the region around Namur in Wallonia (Belgium) in the late 15th century, being introduced to England in 1491. The fuel used in these was invariably charcoal. The successful substitution of coke for charcoal is widely attributed to Abraham Darby in 1709. The efficiency of the process was further enhanced by the practice of preheating the combustion air (hot blast), patented by James Beaumont Neilson in 1828.

## China

An illustration of furnace bellows operated by waterwheels, from the *Nong Shu*, by Wang Zhen, 1313 AD, during the Yuan Dynasty of China

The oldest extant blast furnaces were built during the Han Dynasty of China in the 1st century AD. However, cast iron farm tools and weapons were widespread in China by the 5th century BC, while 3rd century BC iron smelters employed an average workforce of over two hundred men. These early furnaces had clay walls and used phosphorus-containing minerals as a flux. The effectiveness of the Chinese blast furnace was enhanced during this period by the engineer Du Shi (c. 31 AD), who applied the power of waterwheels to piston-bellows in forging cast iron.

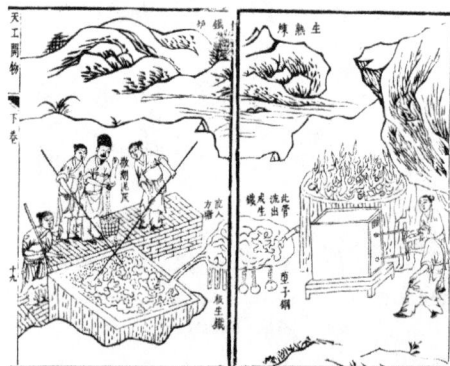

The left picture illustrates the fining process to make wrought iron from pig iron, with the right illustration displaying men working a blast furnace of smelting iron ore producing pig iron, from the *Tiangong Kaiwu* encyclopedia, 1637

While it was long thought that the Chinese had developed the blast furnace and cast iron as their first method of iron production, Donald Wagner (the author of the above referenced study) has published a more recent paper that supersedes some of the statements in the earlier work; the newer paper still places the date of the first cast-iron artifacts at the 5th and 4th centuries BC, but also provides evidence of earlier bloomery furnace use, which migrated from China to Western parts and Central Asia during the beginning of the Chinese Bronze Age of the late Longshan culture (2000 BC). He suggests that early blast furnace and cast iron production evolved from furnaces used to melt bronze. Certainly, though, iron was essential to military success by the time the State of Qin had unified China (221 BC). Usage of the blast and cupola furnace remained widespread during the Song and Tang Dynasties. By the 11th century AD, the Song Dynasty Chinese iron industry made a remarkable switch of resources from charcoal to bituminous coal in casting iron and steel, sparing thousands of acres of woodland from felling. This may have happened as early as the 4th century AD.

The Chinese blast furnace remained in use well until the 20th century. The backyard furnaces favoured by Mao Zedong during the Great Leap Forward were of this type. In the regions with strong traditions of metallurgy, the steel production actually increased during this period. In the regions where there was no tradition of steelmaking or where the ironmasters knowing the traditional skills or the scientific principles of the blast furnace process had been killed, the results were less than satisfactory.

## Elsewhere in the Ancient World

In most places in the world other than in China, there is no evidence of the use of the blast furnace (proper). Instead, iron was made by direct reduction in bloomeries. The bloomery process was invented probably in modern-day Xinjiang or other parts of Western China by Hans or Mongols around 800 BC. Iron finds in China proper are few before bloomeries were supplanted by the **blast furnace** in the 5th century BC which seems to have developed independently in the southern Chinese cultural sphere. An exception would be the Haya people of northwestern Tanzania, who are renowned for creating steel using a blast furnace process and refining process very similar to open hearth process for possibly as long as 2000 years.

In Europe, the Greeks, Celts, Romans, and Carthaginians all used this process. Several examples have been found in France, and materials found in Tunisia suggest they were used there as well as in Antioch (south-central Turkey, between Syria and the Mediterranean Sea) during the Hellenistic Period. Though little is known of it during the Dark Ages, the process probably continued in use. Similarly, smelting in bloomery-type furnaces in West Africa and forging for tools appear in the Nok culture in Africa by 500 BC. The earliest records of bloomery-type furnaces in East Africa are discoveries of smelted iron and carbon in Nubia and Axum that date back between 1,000–500 BCE. Particularly in Meroë, there are known to have been ancient bloomeries that produced metal tools for the Nubians and Kushites and produced surplus for their economy.

Bloomeries have also been discovered and recorded to have been created in medieval West Africa with some of the metalworking Bantu civilizations such as the Bunyoro Empire and the Nyoro people.

## Medieval Europe

### Catalan forges

The simplest forge, known as the Corsican, was used prior to the advent of Christianity. Examples of improved bloomeries are the Stückofen[fr] (sometimes called wolf-furnace) or the Catalan forge, which remained until the beginning of the 19th century. The Catalan forge was invented in Catalonia, Spain, during the 8th century. Instead of using natural draught, air was pumped in by a *trompe*, resulting in better quality iron and an increased capacity. This pumping of airstream in with bellows is known as *cold blast*, and it increases the fuel efficiency of the bloomery and improves yield. The Catalan forges can also be built bigger than natural draught bloomeries.

Modern experimental archaeology and history re-enactment have shown there is only a very short step from the Catalan forge to the true blast furnace, where the iron is gained as pig iron in liquid phase. Usually, obtaining the iron in liquid phase is actually undesired, and the temperature is intentionally kept below the melting point of iron, since while removing the solid bloom mechanically is tedious and means batch process instead of continuous process, it is almost pure iron and can be worked immediately. On the other hand, pig iron is the eutectic mixture of carbon and iron and needs to be decarburized to produce steel or wrought iron, which was extremely tedious in the Middle Ages.

### Oldest European Blast Furnaces

The first blast furnace of Germany as depicted in a miniature in the Deutsches Museum

The oldest known blast furnaces in the West were built in Dürstel in Switzerland, the Märkische Sauerland in Germany, and at Lapphyttan in Sweden, where the complex was active between 1205 and 1300. At Noraskog in the Swedish parish of Järnboås, there have also been found traces of blast furnaces dated even earlier, possibly to around 1100. These early blast furnaces, like the Chinese examples, were very inefficient compared to those used today. The iron from the Lapphyttan complex was used to produce balls of wrought iron known as osmonds, and these were traded internationally – a possible reference occurs in a treaty with Novgorod from 1203 and several certain references in accounts of English customs from the 1250s and 1320s. Other furnaces of the 13th to 15th centuries have been identified in Westphalia.

The technology of blast furnace may have either been transferred from China, or may have been an indigenous innovation. Al-Qazvini in the 13th century and other travellers subsequently noted an iron industry in the Alburz Mountains to the south of the Caspian Sea. This is close to the silk route, so that the use of technology derived from China is conceivable. Much later descriptions record blast furnaces about three metres high. As the Varangian Rus' people from Scandinavia traded with the Caspian (using their Volga trade route, it is possible that the technology reached Sweden by this means. High quality ores, water power for bellows for blast and wood for charcoal are readily obtainable in Sweden. However, since blast furnace has also been invented independently in Africa by the Haya people, it is more likely the process has been invented in Scandinavia independently. The step from bloomery to true blast furnace is not big. Simply just building a bigger furnace and using bigger bellows to increase the volume of the blast and hence the amount of oxygen leads inevitably into higher temperatures, bloom melting into liquid iron and, cast iron flowing from the smelters. Already the Vikings are known to have used double bellows, which greatly increases the volumetric flow of the blast.

This Caspian region may also separately be the technological source for at furnace at Ferriere, described by Filarete. Water-powered bellows at Semogo in northern Italy in 1226 in a two-stage process. In this, the molten iron was tapped twice a day into water thereby granulating it.

## Cistercian Contributions

One means by which certain technological advances were transmitted within Europe was a result of the General Chapter of the Cistercian monks. This may have included the blast furnace, as the Cistercians are known to have been skilled metallurgists. According to Jean Gimpel, their high level of industrial technology facilitated the diffusion of new techniques: "Every monastery had a model factory, often as large as the church and only several feet away, and waterpower drove the machinery of the various industries located on its floor." Iron ore deposits were often donated to the monks along with forges to extract the iron, and within time surpluses were being offered for sale. The Cistercians became the leading iron producers in Champagne, France, from the mid-13th century to the 17th century, also using the phosphate-rich slag from their furnaces as an agricultural fertilizer.

Archaeologists are still discovering the extent of Cistercian technology. At Laskill, an outstation of Rievaulx Abbey and the only medieval blast furnace so far identified in Britain, the slag produced was low in iron content. Slag from other furnaces of the time contained a substantial concentration of iron, whereas Laskill is believed to have produced cast iron quite efficiently. Its date is not yet clear, but it probably did not survive until Henry VIII's Dissolution of the Monasteries in the late 1530s, as an agreement (immediately after that) concerning the "smythes" with the Earl of Rutland in 1541 refers to blooms. Nevertheless, the means by which the blast furnace spread in medieval Europe has not finally been determined.

## Origin and Spread of Early Modern Blast Furnaces

The direct ancestor of these used in France and England was in the Namur region in what is now Wallonia (Belgium). From there, they spread first to the Pays de Bray on the eastern boundary of Normandy and from there to the Weald of Sussex, where the first furnace (called Queenstock) in Buxted was built in about 1491, followed by one at Newbridge in Ashdown Forest in 1496. They remained few in number until about 1530 but many were built in the following decades in the Weald, where the iron industry perhaps reached its peak about 1590. Most of the pig iron from these furnaces was taken to finery forges for the production of bar iron.

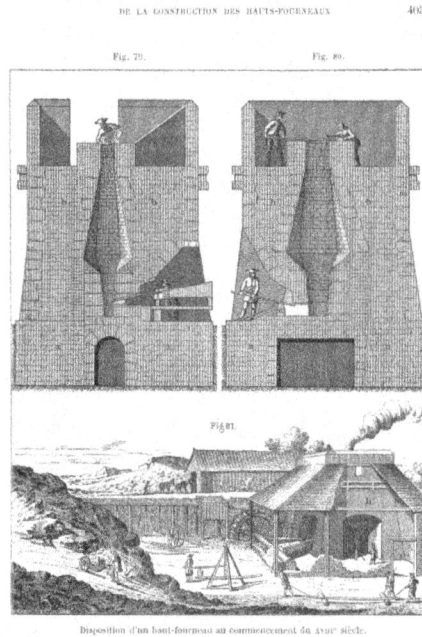

Period drawing of an 18th-century blast furnace

The first British furnaces outside the Weald appeared during the 1550s, and many were built in the remainder of that century and the following ones. The output of the industry probably peaked about 1620, and was followed by a slow decline until the early 18th century. This was apparently because it was more economic to import iron from Sweden and elsewhere than to make it in some more remote British locations. Charcoal that was economically available to the industry was probably being consumed as fast as the wood to make it grew. The Backbarrow blast furnace built in Cumbria in 1711 has been described as the first efficient example.

The first blast furnace in Russia opened in 1637 near Tula and was called the Gorodishche Works. The blast furnace spread from here to the central Russia and then finally to the Urals.

## Coke Blast Furnaces

The original blast furnaces at Blists Hill, Coalbrookdale

Charging the experimental blast furnace, Fixed Nitrogen Research Laboratory, 1930

In 1709, at Coalbrookdale in Shropshire, England, Abraham Darby began to fuel a blast furnace with coke instead of charcoal. Coke's initial advantage was its lower cost, mainly because making coke required much less labor than cutting trees and making charcoal, but using coke also overcame localized shortages of wood, especially in Britain and on the Continent. Metallurgical grade coke will bear heavier weight than charcoal, allowing larger furnaces. A disadvantage is that coke contains more impurities than charcoal, with sulfur being especially detrimental to the iron's quality.

Coke iron was initially only used for foundry work, making pots and other cast iron goods. Foundry work was a minor branch of the industry, but Darby's son built a new furnace at nearby Horsehay, and began to supply the owners of finery forges with coke pig iron for the production of bar iron. Coke pig iron was by this time cheaper to produce than charcoal pig iron. The use of a coal-derived fuel in the iron industry was a key factor in the British Industrial Revolution. Darby's original blast furnace has been archaeologically excavated and can be seen in situ at Coalbrookdale, part of the Ironbridge Gorge Museums. Cast iron from the furnace was used to make girders for the world's first iron bridge in 1779. The Iron Bridge crosses the River Severn at Coalbrookdale and remains in use for pedestrians.

## Hot Blast

Hot Blast was the single most important advance in fuel efficiency of the blast furnace and was one of the most important technologies developed during the Industrial Revolution. Hot blast was patented by James Beaumont Neilson at Wilsontown Ironworks in Scotland in 1828. Within a few years of the introduction, hot blast was developed to the point where fuel consumption was cut by one-third using coke or two-thirds using coal, while furnace capacity was also significantly increased. Within a few decades, the practice was to have a "stove" as large as the furnace next to it into which the waste gas (containing CO) from the furnace was directed and burnt. The resultant heat was used to preheat the air blown into the furnace.

Hot blast enabled the use of raw anthracite coal, which was difficult to light, to the blast furnace. Anthracite was first tried successfully by George Crane at Ynyscedwyn ironworks in south Wales in 1837. It was taken up in America by the Lehigh Crane Iron Company at Catasauqua, Pennsylvania, in 1839.

## Modern Furnaces

### Iron Blast Furnaces

The blast furnace remains an important part of modern iron production. Modern furnaces are highly efficient, including Cowper stoves to pre-heat the blast air and employ recovery systems to extract the heat from the hot gases exiting the furnace. Competition in industry drives higher production rates. The largest blast furnaces have a volume around 5,580 m³ (197,000 cu ft) and can produce around 80,000 tonnes (79,000 long tons; 88,000 short tons) of iron per week.

This is a great increase from the typical 18th-century furnaces, which averaged about 360 tonnes (350 long tons; 400 short tons) per year. Variations of the blast furnace, such as the Swedish electric blast furnace, have been developed in countries which have no native coal resources.

### Lead Blast Furnaces

Blast furnaces are currently rarely used in copper smelting, but modern lead smelting blast furnaces are much shorter than iron blast furnaces and are rectangular in shape. The overall shaft height is around 5 to 6 m. Modern lead blast furnaces are constructed using water-cooled steel or copper jackets for the walls, and have no refractory linings in the side walls. The base of the furnace is a hearth of refractory material (bricks or castable refractory). Lead blast furnaces are often open-topped rather than having the charging bell used in iron blast furnaces.

The blast furnace used at the Nyrstar Port Pirie lead smelter differs from most other lead blast furnaces in that it has a double row of tuyeres rather than the single row normally used. The lower shaft of the furnace has a chair shape with the lower part of the shaft being narrower than the upper. The lower row of tuyeres being located in the narrow part of the shaft. This allows the upper part of the shaft to be wider than the standard.

### Zinc Blast Furnaces (Imperial Smelting Furnaces)

The blast furnaces used in the Imperial Smelting Process ("ISP") were developed from the standard lead blast furnace, but are fully sealed. This is because the zinc produced by these furnaces is recovered as metal from the vapor phase, and the presence of oxygen in the off-gas would result in the formation of zinc oxide.

Blast furnaces used in the ISP have a more intense operation than standard lead blast furnaces, with higher air blast rates per m² of hearth area and a higher coke consumption.

Zinc production with the ISP is more expensive than with electrolytic zinc plants, so several smelters operating this technology have closed in recent years. However, ISP furnaces have the advantage of being able to treat zinc concentrates containing higher levels of lead than can electrolytic zinc plants.

## Modern Process

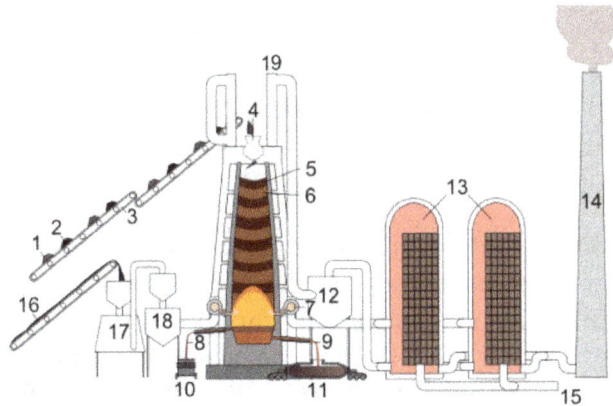

Blast furnace placed in an installation

1. Iron ore + limestone sinter, 2. Coke, 3. Elevator, 4. Feedstock inlet, 5. Layer of coke, 6. Layer of sinter pellets of ore and limestone, 7. Hot blast (around 1200 °C), 8. Removal of slag, 9. Tapping of molten pig iron, 10. Slag pot, 11. Torpedo car for pig iron, 12. Dust cyclone for separation of solid particles, 13. Cowper stoves for hot blast, 14. Smoke outlet (can be redirected to carbon capture & storage (CCS) tank), 15: Feed air for Cowper stoves (air pre-heaters), 16. Powdered coal, 17. Coke oven, 18. Coke, 19. Blast furnace gas downcomer

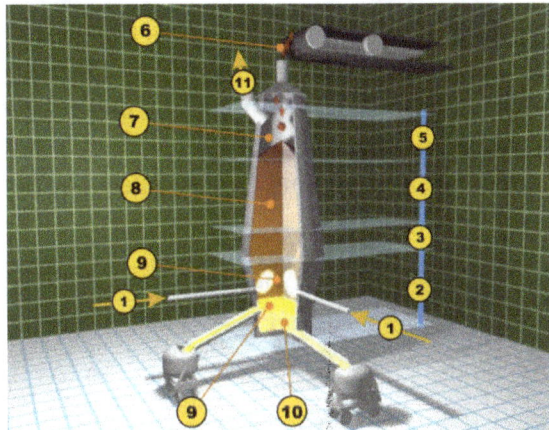

Blast Furnace Diagram

1. Hot blast from Cowper stoves, 2. Melting zone (bosh), 3. Reduction zone of ferrous oxide (barrel), 4. Reduction zone of ferric oxide (stack), 5. Pre-heating zone (throat), 6. Feed of ore, limestone, and coke, 7. Exhaust gases, 8. Column of ore, coke and limestone, 9. Removal of slag, 10. Tapping of molten pig iron, 11. Collection of waste gases

Modern furnaces are equipped with an array of supporting facilities to increase efficiency, such as ore storage yards where barges are unloaded. The raw materials are transferred to the stockhouse complex by ore bridges, or rail hoppers and ore transfer cars. Rail-mounted scale cars or computer controlled weight hoppers weigh out the various raw materials to yield the desired hot metal and slag chemistry. The raw materials are brought to the top of the blast furnace via a skip car powered by winches or conveyor belts.

There are different ways in which the raw materials are charged into the blast furnace. Some blast furnaces use a "double bell" system where two "bells" are used to control the entry of raw material into the blast furnace. The purpose of the two bells is to minimize the loss of hot gases in the blast furnace. First, the raw materials are emptied into the upper or small bell which then opens to empty the charge into the large bell. The small bell then closes, to seal the blast furnace, while the large bell rotates to provide specific distribution of materials before dispensing the charge into the blast furnace. A more recent design is to use a "bell-less" system. These systems use multiple hoppers to contain each raw material, which is then discharged into the blast furnace through valves. These valves are more accurate at controlling how much of each constituent is added, as compared to the skip or conveyor system, thereby increasing the efficiency of the furnace. Some of these bell-less systems also implement a discharge chute in the throat of the furnace (as with the Paul Wurth top) in order to precisely control where the charge is placed.

The iron making blast furnace itself is built in the form of a tall structure, lined with refractory brick, and profiled to allow for expansion of the charged materials as they heat during their descent, and subsequent reduction in size as melting starts to occur. Coke, limestone flux, and iron ore (iron oxide) are charged into the top of the furnace in a precise filling order which helps control gas flow and the chemical reactions inside the furnace. Four "uptakes" allow the hot, dirty gas high in carbon monoxide content to exit the furnace throat, while "bleeder valves" protect the top of the furnace from sudden gas pressure surges. The coarse particles in the exhaust gas settle in the "dust catcher" and are dumped into a railroad car or truck for disposal, while the gas itself flows through a venturi scrubber and/or electrostatic precipitators and a gas cooler to reduce the temperature of the cleaned gas.

The "casthouse" at the bottom half of the furnace contains the bustle pipe, water cooled copper tuyeres and the equipment for casting the liquid iron and slag. Once a "taphole" is drilled through the refractory clay plug, liquid iron and slag flow down a trough through a "skimmer" opening, separating the iron and slag. Modern, larger blast furnaces may have as many as four tapholes and two casthouses. Once the pig iron and slag has been tapped, the taphole is again plugged with refractory clay.

Tuyeres of Blast Furnace at Gerdau, India

The tuyeres are used to implement a hot blast, which is used to increase the efficiency of the blast furnace. The hot blast is directed into the furnace through water-cooled copper nozzles called tuyeres near the base. The hot blast temperature can be from 900 °C to 1300 °C (1600 °F to 2300 °F) depending on the stove design and condition. The temperatures they deal with may be 2000 °C to 2300 °C (3600 °F to 4200 °F). Oil, tar, natural gas, powdered coal and oxygen can also be injected into the furnace at tuyere level to combine with the coke to release additional energy and increase the percentage of reducing gases present which is necessary to increase productivity.

## Process Engineering and Chemistry

Blast furnaces of Třinec Iron and Steel Works, Czech Republic

Blast furnaces operate on the principle of chemical reduction whereby carbon monoxide, having a stronger affinity for the oxygen in iron ore than iron does, reduces the iron to its elemental form. Blast furnaces differ from bloomeries and reverberatory furnaces in that in a blast furnace, flue gas is in direct contact with the ore and iron, allowing carbon monoxide to diffuse into the ore and reduce the iron oxide to elemental iron mixed with carbon. The blast furnaces operates as a countercurrent exchange process whereas a bloomery does not. Another difference is that bloomeries operate as a batch process while blast furnaces operate continuously for long periods because they are difficult to start up and shut down. Also, the carbon in pig iron lowers the melting point below that of steel or pure iron; in contrast, iron does not melt in a bloomery.

Carbon monoxide also reduces silica which has to be removed from the pig iron. The silica is reacted with calcium oxide (burned limestone) and forms a slag which floats to the surface of the molten pig iron. The direct contact of flue gas with the iron causes contamination with sulfur if it is present in the fuel. Historically, to prevent contamination from sulfur, the best quality iron was produced with charcoal.

The downward moving column of ore, flux, coke or charcoal and reaction products must be porous enough for the flue gas to pass through. This requires the coke or charcoal to be in large enough particles to be permeable, meaning there cannot be an excess of fine particles. Therefore, the coke must be strong enough so it will not be crushed by the weight of the material above it. Besides physical strength of the coke, it must also be low in sulfur, phosphorus, and ash. This necessitates the use of metallurgical coal, which is a premium grade due to its relative scarcity.

The main chemical reaction producing the molten iron is:

$$Fe_2O_3 + 3CO \rightarrow 2Fe + 3CO_2$$

This reaction might be divided into multiple steps, with the first being that preheated blast air blown into the furnace reacts with the carbon in the form of coke to produce carbon monoxide and heat:

$$2\,C_{(s)} + O_{2(g)} \rightarrow 2\,CO_{(g)}$$

The hot carbon monoxide is the reducing agent for the iron ore and reacts with the iron oxide to produce molten iron and carbon dioxide. Depending on the temperature in the different parts of the furnace (warmest at the bottom) the iron is reduced in several steps. At the top, where the temperature usually is in the range between 200 °C and 700 °C, the iron oxide is partially reduced to iron(II,III) oxide, $Fe_3O_4$.

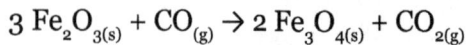

$$3\ Fe_2O_{3(s)} + CO_{(g)} \rightarrow 2\ Fe_3O_{4(s)} + CO_{2(g)}$$

At temperatures around 850 °C, further down in the furnace, the iron(II,III) is reduced further to iron(II) oxide:

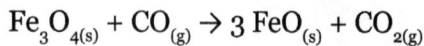

$$Fe_3O_{4(s)} + CO_{(g)} \rightarrow 3\ FeO_{(s)} + CO_{2(g)}$$

Hot carbon dioxide, unreacted carbon monoxide, and nitrogen from the air pass up through the furnace as fresh feed material travels down into the reaction zone. As the material travels downward, the counter-current gases both preheat the feed charge and decompose the limestone to calcium oxide and carbon dioxide:

$$CaCO_{3(s)} \rightarrow CaO_{(s)} + CO_{2(g)}$$

The calcium oxide formed by decomposition reacts with various acidic impurities in the iron (notably silica), to form a fayalitic slag which is essentially calcium silicate, $CaSiO_3$:

$$SiO_2 + CaO \rightarrow CaSiO_3$$

As the iron(II) oxide moves down to the area with higher temperatures, ranging up to 1200 °C degrees, it is reduced further to iron metal:

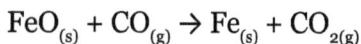

$$FeO_{(s)} + CO_{(g)} \rightarrow Fe_{(s)} + CO_{2(g)}$$

The carbon dioxide formed in this process is re-reduced to carbon monoxide by the coke:

$$C_{(s)} + CO_{2(g)} \rightarrow 2\ CO_{(g)}$$

The temperature-dependent equilibrium controlling the gas atmosphere in the furnace is called the Boudouard reaction:

$$2CO \rightarrow CO_2 + C$$

The "pig iron" produced by the blast furnace has a relatively high carbon content of around 4–5%, making it very brittle, and of limited immediate commercial use. Some pig iron is used to make cast iron. The majority of pig iron produced by blast furnaces undergoes further processing to reduce the carbon content and produce various grades of steel used for construction materials, automobiles, ships and machinery.

Although the efficiency of blast furnaces is constantly evolving, the chemical process inside the blast furnace remains the same. According to the American Iron and Steel Institute: "Blast furnaces will survive into the next millennium because the larger, efficient furnaces can produce hot metal at costs competitive with other iron making technologies." One of the biggest drawbacks of the blast furnaces is the inevitable carbon dioxide production as iron is reduced from iron oxides by carbon and as of 2016, there is no economical substitute – steelmaking is one of the largest industrial contributors of the $CO_2$ emissions in the world.

The challenge set by the greenhouse gas emissions of the blast furnace is being addressed in an ongoing European Program called ULCOS (Ultra Low $CO_2$ Steelmaking). Several new process routes have been proposed and investigated in depth to cut specific emissions ($CO_2$ per ton of steel) by at least 50%. Some rely on the capture and further storage (CCS) of $CO_2$, while others choose decarbonizing iron and steel production, by turning to hydrogen, electricity and biomass. In the nearer term, a technology that incorporates CCS into the blast furnace process itself and is called the Top-Gas Recycling Blast Furnace is under development, with a scale-up to a commercial size blast furnace under way. The technology should be fully demonstrated by the end of the 2010s, in line with the timeline set, for example, by the EU to cut emissions significantly. Broad deployment could take place from 2020 on.

## Manufacture of Stone Wool

Stone wool or rock wool is a spun mineral fibre used as an insulation product and in hydroponics. It is manufactured in a blast furnace fed with diabase rock which contains very low levels of metal oxides. The resultant slag is drawn off and spun to form the rock wool product. Very small amounts of metals are also produced which are an unwanted by-product and run to waste.

## Decommissioned Blast Furnaces as Museum Sites

For a long time, it was normal procedure for a decommissioned blast furnace to be demolished and either be replaced with a newer, improved one, or to have the entire site demolished to make room for follow-up use of the area. In recent decades, several countries have realized the value of blast furnaces as a part of their industrial history. Rather than being demolished, abandoned steel mills were turned into museums or integrated into multi-purpose parks. The largest number of preserved historic blast furnaces exists in Germany; other such sites exist in Spain, France, the Czech Republic, Japan, Luxembourg, Poland, Romania, Mexico, Russia and the United States.

## Cold Blast

Cold blast, in ironmaking, refers to a furnace where air is not preheated before being blown into the furnace. This represents the earliest stage in the development of ironmaking. Until the 1820s, the use of cold air was thought to be preferable to hot air for the production of high-quality iron; this effect was due to the reduced moisture in cool winter air.

The discovery by James Beaumont Neilson in about 1825 of the beneficial effects of the hot blast led to the rapid obsolescence of cold blast ironworks in Great Britain, where hot blast was in general use by 1835. Cold blast ironworks survived longer in the United States, but the use of hot blast as a method of smelting iron with anthracite was introduced in 1836, and the increasing US production of coke gradually drove out the cold blast furnaces. However, one of the last known operating cold blast charcoal furnaces, Pleasant (formerly Eagle) Furnace, in Curtin, Pennsylvania did not close until 1921.

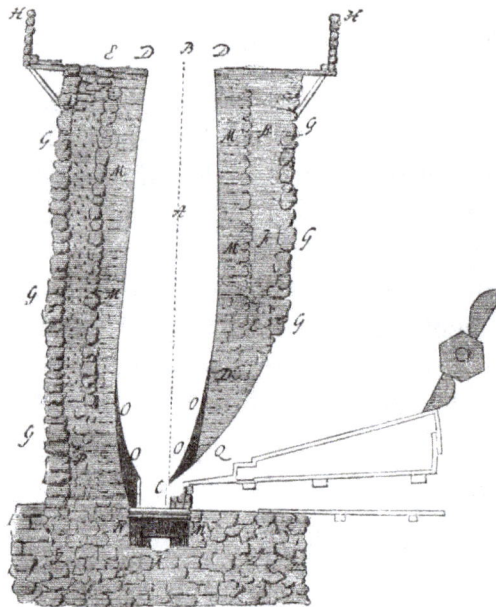

Section of a 17th and 18th century blast furnace. The bellows, at right, draw in air directly from the atmosphere.

## Hot Blast

Hot blast refers to the preheating of air blown into a blast furnace or other metallurgical process. As first developed it worked by alternately storing heat from the furnace flue gas in a firebrick lined vessel with multiple chambers, then blowing combustion air through the hot chamber. This is known as regenerative heating. This has the result of considerably reducing the fuel consumed in the process. Hot blast was invented and patented for iron furnaces by James Beaumont Neilson in 1828 at Wilsontown Ironworks in Scotland, but was later applied in other contexts, including late bloomeries. Later the carbon monoxide in the flue gas was burned to provide additional heat.

Blast furnace (left), and three Cowper stoves (right) used to preheat the air blown into the furnace

Hot blast furnace: note the flow of air from the two stoves in the background to the blast furnace, and hot air from the furnace being drawn off to heat the stoves

Hot blast was the single most important advance in fuel efficiency of the blast furnace and was one of the most important technologies developed during the Industrial Revolution. Hot blast also allowed higher furnace temperatures, which increased the capacity of furnaces.

## History

### Invention and Spread

James Beaumont Neilson, previously foreman at Glasgow gas works invented the system of pre-heating the blast for a furnace. He found that by increasing the temperature of the incoming air to 300 degrees Fahrenheit, he could reduce the fuel consumption from 8.06 tons to 5.16 tons with further reductions at even higher temperatures. He with partners including Charles Macintosh patented this in 1828. Initially the heating vessel was made of wrought iron plates, but these oxidized, and he substituted a cast iron vessel.

The spread of this technology to other parts of Britain was relatively slow. However, by 1840, 58 ironmasters had taken out licenses, yielding a royalty income of £30,000 per year for Neilson and his partners. However they had to engage in substantial litigation, ultimately successful, over the following years to enforce the patent against infringers. By the time the patent expired there were 80 licenses. In 1843, just after it expired, 42 of the 80 furnaces in south Staffordshire were using hot blast, and uptake in south Wales was even slower.

Other advantages in using hot blast were that raw coal could be used instead of coke and in Scotland, the relatively poor "black band" ironstone could be profitably smelted. It also increased the daily output of each furnace, in the case of Calder ironworks from 5.6 tons per day in 1828 to 8.2 in 1833, which made Scotland the lowest cost steel producing region in Britain in the 1830s.

Nevertheless, early hot blast stoves were troublesome, as thermal expansion and contraction were liable to cause breakages in the pipes. This was to some extent remedied by supporting the pipes on rollers. It was also necessary to devise new methods of connecting the blast pipes to the tuyeres, as leather could not longer be used for making the connection.

Ultimately this principle was applied even more efficiently in regenerative heat exchangers, such as the Cowper stove (which preheat the incoming blast air using waste heat from the flue gas and are used in blast furnaces to this day), and in the open hearth furnace (for making steel) by the Siemens-Martin process.

Independently, George Crane and David Thomas, of the Yniscedwyn Works in Wales, conceived of the same idea, and Crane filed for a British patent in 1836. They began producing iron by the new process on February 5, 1837. Crane subsequently bought Gessenhainer's patent and patented additions to it, controlling the use of the process in both Britain and the U.S. While Crane remained in Wales, Thomas would move to the U.S. on behalf of the Lehigh Coal and Navigation Company and found the Lehigh Crane Iron Company to make use of the process.

## Anthracite in Ironmaking

Hot blast allowed the use of anthracite in iron smelting. It also allowed use of lower quality coal because less fuel meant proportionately less sulfur and ash.

At the time the process was invented, good coking coal was only available in sufficient quantities in Great Britain and western Germany, so iron furnaces in the U.S. were using charcoal. This meant that any given iron furnace required vast tracts of forested land for charcoal production, and generally went out of blast when the nearby woods had been felled. Attempts to use anthracite as a fuel had all ended in failure, as the coal resisted ignition under cold blast conditions. In 1831, Dr. Frederick W. Gessenhainer filed for a U.S. patent on the use of hot blast and anthracite to smelt iron. He produced a small quantity of anthracite iron by this method at Valley Furnace near Pottsville, Pennsylvania in 1836, but due to breakdowns and his illness and death in 1838, he was not able to carry the process into large-scale production.

Anthracite was displaced by coke in the U.S. after the Civil War. Coke was more porous and able to support the heavier loads in the vastly larger furnaces of the late 19th century.

## Steel

For steel the hot blast temperature can be from 900 °C to 1300 °C (1600 °F to 2300 °F) depending on the stove design and condition. The temperatures they deal with may be 2000 °C to 2300 °C (3600 °F to 4200 °F). Oil, tar, natural gas, powdered coal and oxygen can also be injected into the furnace at tuyere level to combine with the coke to release additional energy which is necessary to increase productivity.

## References

- Peter J. Golas (25 February 1999). Science and Civilisation in China: Volume 5, Chemistry and Chemical Technology, Part 13, Mining. Cambridge University Press. p. 152. ISBN 978-0-521-58000-7.

- Needham, Joseph (1986), Science and Civilisation in China, Volume 4: Physics and Physical Technology, Part 2, Mechanical Engineering, Taipei: Cambridge University Press, p. 370, ISBN 0-521-05803-1

- Robert O. Collins; James McDonald Burns (2007). A History of Sub-Saharan Africa. Cambridge University Press. p. 61. ISBN 978-0-521-86746-7. Retrieved 12 July 2012.

- David N. Edwards (2004). The Nubian Past: An Archaeology of the Sudan. Psychology Press. p. 173. ISBN 978-0-415-36987-9. Retrieved 12 July 2012.

- Eugenia W. Herbert (1993). Iron, Gender, and Power: Rituals of Transformation in African Societies. Indiana University Press. p. 102. ISBN 978-0-253-20833-0. Retrieved 12 July 2012.

- Julius H. Strassburger (1969). Blast Furnace-theory and Practice. Gordon and Breach Science Publishers. p. 4. ISBN 978-0-677-10420-1. Retrieved 12 July 2012.

- Rosen, William (2012). The Most Powerful Idea in the World: A Story of Steam, Industry and Invention. University Of Chicago Press. p. 149. ISBN 978-0226726342.

- Landes, David. S. (1969). The Unbound Prometheus: Technological Change and Industrial Development in Western Europe from 1750 to the Present. Cambridge, New York: Press Syndicate of the University of Cambridge. p. 92. ISBN 0-521-09418-6.

- Rayner-Canham & Overton (2006), Descriptive Inorganic Chemistry, Fourth Edition, New York: W. H. Freeman and Company, pp. 534–535, ISBN 978-0-7167-7695-6

- Landes, David. S. (1969). The Unbound Prometheus: Technological Change and Industrial Development in Western Europe from 1750 to the Present. Cambridge, New York: Press Syndicate of the University of Cambridge. p. 92. ISBN 0-521-09418-6.

- Bartholomew, Craig L.; Metz, Lance E. (1988). Bartholomew, Ann (ed.), ed. The Anthracite Industry of the Lehigh Valley. Center for Canal History and Technology. ISBN 0-930973-08-9.

- Rosenberg, Nathan (1982). Inside the Black Box: Technology and Economics. Cambridge, New York: Cambridge University Press. p. 88. ISBN 0-521-27367-6.

# 6

# Understanding of Steel Making

Steel is an alloy of iron that has high tensile strength and can be produced at low costs. This chapter describes the modern steel making process. A section of the chapter illustrates the workings of a steel mill and its functions. The chapter strategically encompasses and incorporates the major components and key concepts of steelmaking, providing a complete understanding.

## Steelmaking

Steelmaking is the process for producing steel from iron ore and scrap. In steelmaking, impurities such as nitrogen, silicon, phosphorus, sulfur and excess carbon are removed from the raw iron, and alloying elements such as manganese, nickel, chromium and vanadium are added to produce different grades of steel. Limiting dissolved gases such as nitrogen and oxygen, and entrained impurities (termed "inclusions") in the steel is also important to ensure the quality of the products cast from the liquid steel.

Steel mill with two arc furnaces

Steelmaking has existed for millennia, but it was not commercialized until the 19th century. The ancient craft process of steelmaking was the crucible process. In the 1850s and 1860s, the Bessemer process and the Siemens-Martin process turned steelmaking into a heavy industry. Today there are two major commercial processes for making steel, namely basic oxygen steelmaking, which has liquid pig-iron from the blast furnace and scrap steel as the main feed materials, and electric arc furnace (EAF) steelmaking, which uses scrap steel or direct reduced iron (DRI) as the main feed materials. Oxygen steelmaking is fuelled predominantly by the exothermic nature of the reactions inside the vessel where as in EAF steelmaking, electrical energy is used to melt the solid scrap and/or DRI materials. In recent times, EAF steelmaking technology has evolved closer to oxygen steelmaking as more chemical energy is introduced into the process.

# History

Bethlehem Steel (Bethlehem, Pennsylvania facility pictured) was one of the world's largest manufacturers of steel before its 2003 closure.

Steelmaking has played a crucial role development of modern technological societies. Cast iron is a hard brittle material that is difficult to work, whereas steel is malleable, relatively easily formed and a versatile material. For much of human history, steel has only been made in small quantities. Since the invention of the Bessemer process in the 19th century and subsequent technological developments in injection technology and process control, mass production of steel has become an integral part of the world's economy and a key indicator of technological development. The earliest means of producing steel was in a bloomery.

Early modern methods of producing steel were often labour-intensive and highly skilled arts.

- finery forge, in which the German finery process could be managed to produce steel.

- blister steel and crucible steel.

An important aspect of the Industrial Revolution was the development of large-scale methods of producing forgeable metal (bar iron or steel). The puddling furnace was initially a means of producing wrought iron, but was later applied to steel production.

The real revolution in steelmaking only began at the end of the 1850s when the Bessemer process became the first successful method of steelmaking in quantity, followed by the open-hearth furnace.

## Modern Processes

Modern steelmaking processes can be broken into two categories: primary and secondary steelmaking. Primary steelmaking involves converting liquid iron from a blast furnace and steel scrap into steel via basic oxygen steelmaking or melting scrap steel and/or direct reduced iron (DRI) in an electric arc furnace. Secondary steelmaking involves refining of the crude steel before casting and the various operations are normally carried out in ladles. In secondary metallurgy, alloying agents are added, dissolved gases in the steel are lowered, and inclusions are removed or altered chemically to ensure that high-quality steel is produced after casting.

## Primary Steelmaking

Basic oxygen steelmaking is a method of primary steelmaking in which carbon-rich molten pig iron is made into steel. Blowing oxygen through molten pig iron lowers the carbon content of the alloy and changes it into steel. The process is known as *basic* due to the chemical nature of the refractories—calcium oxide and magnesium oxide—that line the vessel to withstand the high temperature and corrosive nature of the molten metal and slag in the vessel. The slag chemistry of the process is also controlled to ensure that impurities such as silicon and phosphorus are removed from the metal.

The process was developed in 1948 by Robert Durrer and commercialized in 1952–53 by Austrian VOEST and ÖAMG. The LD converter, named after the Austrian towns of Linz and Donawitz (a district of Leoben) is a refined version of the Bessemer converter where blowing of air is replaced with blowing oxygen. It reduced capital cost of the plants, time of smelting, and increased labor productivity. Between 1920 and 2000, labour requirements in the industry decreased by a factor of 1,000, from more than 3 worker-hours per tonne to just 0.003. The vast majority of steel manufactured in the world is produced using the basic oxygen furnace; in 2011, it accounted for 70% of global steel output. Modern furnaces will take a charge of iron of up to 350 tons and convert it into steel in less than 40 minutes, compared to 10–12 hours in an open hearth furnace.

Electric arc furnace steelmaking is the manufacture of steel from scrap or direct reduced iron melted by electric arcs. In an electric arc furnace, a batch of steel ("heat") may be started by loading scrap or direct reduced iron into the furnace, sometimes with a "hot heel" (molten steel from a previous heat). Gas burners may be used to assist with the melt down of the scrap pile in the furnace. As in basic oxygen steelmaking, fluxes are also added to protect the lining of the vessel and help improve the removal of impurities. Electric arc furnace steelmaking typically uses furnaces of capacity around 100 tonnes that produce steel every 40 to 50 minutes for further processing.

By-product gases from the steel making process can be used to generate electricity through the use of reciprocating gas engines.

## Secondary Steelmaking

Secondary steelmaking is most commonly performed in ladles and often referred to as ladle (metallurgy). Some of the operations performed in ladles include de-oxidation (or "killing"), vacuum degassing, alloy addition, inclusion removal, inclusion chemistry modification, de-sulphurisation and homogenisation. It is now common to perform ladle metallurgical operations in gas stirred ladles with electric arc heating in the lid of the furnace. Tight control of ladle metallurgy is associated with producing high grades of steel in which the tolerances in chemistry and consistency are narrow.

## HIsarna Steelmaking

The HIsarna steelmaking process is a process for primary steelmaking in which iron ore is processed almost directly into steel. The process is based around a new type of blast furnace called a

*Cyclone Converter Furnace*, which makes it possible to skip the process of manufacturing pig iron pellets that is necessary for the basic oxygen steelmaking process. Without the necessity for this preparatory step the HIsarna process is more energy-efficient and has a lower carbon footprint than traditional steelmaking processes.

## Steel Mill

Integrated steel mill in the Netherlands. The two large towers are blast furnaces.

A steel mill or steelworks is an industrial plant for the manufacture of steel.

Steel mill can refer to the steel works making rolled products from iron ore, but it also designs, more precisely the plant where steel semi-finished casting products (blooms, ingots, slabs, billets) are made, from molten pig iron or from scraps.

### History

Since the invention of the Bessemer process, steel mills have replaced ironwork, based on puddling or fining methods. New ways to produce steel appeared later: from scraps melted in an electric arc furnace and, more recently, from direct reduced iron processes.

In the late 19th and early 20th centuries the world's largest steel mill was the Barrow Hematite Steel Company steelworks located in Barrow-in-Furriness, United Kingdom. Today, the world's largest steel mill is in Gwangyang, South Korea.

### Integrated Mill

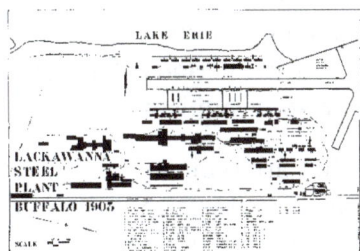

Plan of the Blacksnake Steel Plant ca. 1903, showing the various elements of an integrated steel mill

Blast furnaces of Třinec Iron and Steel Works

Interior of a steel mill

An integrated steel mill has all the functions for primary steel production:

- iron making (conversion of ore to liquid iron),

- steel making (conversion of pig iron to liquid steel),

- casting (solidification of the liquid steel),

- roughing rolling/billet rolling (reducing size of blocks)

- product rolling (finished shapes).

The principal raw materials for an integrated mill are iron ore, limestone, and coal (or coke). These materials are charged in batches into a blast furnace where the iron compounds in the ore give up excess oxygen and become liquid iron. At intervals of a few hours, the accumulated liquid iron is tapped from the blast furnace and either cast into pig iron or directed to other vessels for further steel making operations. Historically the Bessemer process was a major advancement in the production of economical steel, but it has now been entirely replaced by other processes such as the basic oxygen furnace.

Molten steel is cast into large blocks called *blooms*. During the casting process various methods are used, such as addition of aluminum, so that impurities in the steel float to the surface where they can be cut off the finished bloom.

Because of the energy cost and structural stress associated with heating and cooling a blast furnace, typically these primary steel making vessels will operate on a continuous production campaign of several years duration. Even during periods of low steel demand, it may not be feasible to let the blast furnace grow cold, though some adjustment of the production rate is possible.

Integrated mills are large facilities that are typically only economical to build in 2,000,000-ton per year annual capacity and up. Final products made by an integrated plant are usually large structural sections, heavy plate, strip, wire rod, railway rails, and occasionally long products such as bars and pipe.

A major environmental hazard associated with integrated steel mills is the pollution produced in the manufacture of coke, which is an essential intermediate product in the reduction of iron ore in a blast furnace.

Integrated mills may also adopt some of the processes used in mini-mills, such as arc furnaces and direct casting, to reduce production costs.

World integrated steel production capacity is at or close to world demand, so competition between suppliers results in only the most efficient producers remaining viable. However, due to the large employment of integrated plants, often governments will financially assist an obsolescent facility rather than take the risk of having thousands of workers thrown out of jobs.

## Minimill

An ingot of steel entering a rolling mill

A minimill is traditionally a secondary steel producer; however, Nucor (one of the world's largest steel producers), as well as one of its competitors, Commercial Metals Company (CMC) use minimills exclusively. Usually it obtains most of its iron from scrap steel, recycled from used automobiles and equipment or byproducts of manufacturing. Direct reduced iron (DRI) is sometimes used with scrap, to help maintain desired chemistry of the steel, though usually DRI is too expensive to use as the primary raw steelmaking material. A typical mini-mill will have an electric arc furnace for scrap melting, a ladle furnace or vacuum furnace for precision control of chemistry, a strip or billet continuous caster for converting molten steel to solid form, a reheat furnace and a rolling mill.

Originally the mini mill was adapted to production of bar products only, such as concrete reinforcing bar, flats, angles, channels, pipe, and light rails. Since the late 1980s, successful introduction of the direct strip casting process has made mini mill production of strip feasible. Often a mini mill will be constructed in an area with no other steel production, to take advantage of local resources and lower-cost labour. Mini mill plants may specialize, for example, making coils of rod for wire-drawing use, or pipe, or in special sections for transportation and agriculture.

Capacities of mini mills vary; some plants may make as much as 3,000,000 tons per year, a typical size is in the range 200,000 to 400,000 tons per year, and some old or specialty plants may make as little as 50,000 tons per year of finished product. Nucor Corporation, for example, annually produces around 9,100,000 tons of sheet steel from its four sheet mills, 6,700,000 tons of bar steel from its 10 bar mills and 2,100,000 tons of plate steel from its two plate mills.

Since the electric arc furnace can be easily started and stopped on a regular basis, mini mills can follow the market demand for their products easily, operating on 24-hour schedules when demand is high and cutting back production when sales are lower.

# 7

# Corrosion: A Natural Process

Corrosion occurs when metals like iron are exposed to the atmosphere. The metals form stable compounds by chemical and/or electrochemical reactions. The chapter describes in detail the process of anaerobic corrosion which takes place in the presence of anoxic water. The content also talks about cyclic corrosion testing, a test under which samples are subject in laboratories to varying climates that they might encounter in the real world.

## Corrosion

Corrosion is a natural process, which converts a refined metal to a more stable form, such as its oxide, hydroxide, or sulfide. It is the gradual destruction of materials (usually metals) by chemical and/or electrochemical reaction with their environment. Corrosion engineering is the field dedicated to controlling and stopping corrosion.

Rust, the most familiar example of corrosion.

In the most common use of the word, this means electrochemical oxidation of metal in reaction with an oxidant such as oxygen or sulfur. Rusting, the formation of iron oxides, is a well-known example of electrochemical corrosion. This type of damage typically produces oxide(s) or salt(s) of the original metal, and results in a distinctive orange colouration. Corrosion can also occur in materials other than metals, such as ceramics or polymers, although in this context, the term "degradation" is more common. Corrosion degrades the useful properties of materials and structures including strength, appearance and permeability to liquids and gases.

Volcanic gases have accelerated the corrosion of this abandoned mining machinery.

Corrosion on exposed metal.

Many structural alloys corrode merely from exposure to moisture in air, but the process can be strongly affected by exposure to certain substances. Corrosion can be concentrated locally to form a pit or crack, or it can extend across a wide area more or less uniformly corroding the surface. Because corrosion is a diffusion-controlled process, it occurs on exposed surfaces. As a result, methods to reduce the activity of the exposed surface, such as passivation and chromate conversion, can increase a material's corrosion resistance. However, some corrosion mechanisms are less visible and less predictable.

## Galvanic Corrosion

Galvanic corrosion occurs when two different metals have physical or electrical contact with each other and are immersed in a common electrolyte, or when the same metal is exposed to electrolyte with different concentrations. In a galvanic couple, the more active metal (the anode) corrodes at an accelerated rate and the more noble metal (the cathode) corrodes at a slower rate. When immersed separately, each metal corrodes at its own rate. What type of metal(s) to use is readily determined by following the galvanic series. For example, zinc is often used as a sacrificial anode for steel structures. Galvanic corrosion is of major interest to the marine industry and also anywhere water (containing salts) contacts pipes or metal structures.

Galvanic corrosion of aluminium. A 5-mm-thick aluminium alloy plate is physically (and hence, electrically) connected to a 10-mm-thick mild steel structural support. Galvanic corrosion occurred on the aluminium plate along the joint with the steel. Perforation of aluminium plate occurred within 2 years.

Factors such as relative size of anode, types of metal, and operating conditions (temperature, humidity, salinity, etc.) affect galvanic corrosion. The surface area ratio of the anode and cathode directly affects the corrosion rates of the materials. Galvanic corrosion is often prevented by the use of sacrificial anodes.

## Galvanic Series

In any given environment (one standard medium is aerated, room-temperature seawater), one metal will be either more *noble* or more *active* than others, based on how strongly its ions are bound to the surface. Two metals in electrical contact share the same electrons, so that the "tug-of-war" at each surface is analogous to competition for free electrons between the two materials. Using the electrolyte as a host for the flow of ions in the same direction, the noble metal will take electrons from the active one. The resulting mass flow or electric current can be measured to establish a hierarchy of materials in the medium of interest. This hierarchy is called a *galvanic series* and is useful in predicting and understanding corrosion.

## Corrosion Removal

Often it is possible to chemically remove the products of corrosion. For example, phosphoric acid in the form of naval jelly is often applied to ferrous tools or surfaces to remove rust. Corrosion removal should not be confused with electropolishing, which removes some layers of the underlying metal to make a smooth surface. For example, phosphoric acid may also be used to electropolish copper but it does this by removing copper, not the products of copper corrosion.

## Resistance to Corrosion

Some metals are more intrinsically resistant to corrosion than others. There are various ways of protecting metals from corrosion (oxidation) including painting, hot dip galvanizing, and combinations of these.

## Intrinsic Chemistry

The materials most resistant to corrosion are those for which corrosion is thermodynamically unfavorable. Any corrosion products of gold or platinum tend to decompose spontaneously into pure metal, which is why these elements can be found in metallic form on Earth and have long been valued. More common "base" metals can only be protected by more temporary means.

Gold nuggets do not naturally corrode, even on a geological time scale.

Some metals have naturally slow reaction kinetics, even though their corrosion is thermodynamically favorable. These include such metals as zinc, magnesium, and cadmium. While corrosion of these metals is continuous and ongoing, it happens at an acceptably slow rate. An extreme example is graphite, which releases large amounts of energy upon oxidation, but has such slow kinetics that it is effectively immune to electrochemical corrosion under normal conditions.

## Passivation

Passivation refers to the spontaneous formation of an ultrathin film of corrosion products, known as a passive film, on the metal's surface that act as a barrier to further oxidation. The chemical composition and microstructure of a passive film are different from the underlying metal. Typical passive film thickness on aluminium, stainless steels, and alloys is within 10 nanometers. The passive film is different from oxide layers that are formed upon heating and are in the micrometer thickness range – the passive film recovers if removed or damaged whereas the oxide layer does not. Passivation in natural environments such as air, water and soil at moderate pH is seen in such materials as aluminium, stainless steel, titanium, and silicon.

Passivation is primarily determined by metallurgical and environmental factors. The effect of pH is summarized using Pourbaix diagrams, but many other factors are influential. Some conditions that inhibit passivation include high pH for aluminium and zinc, low pH or the presence of chloride ions for stainless steel, high temperature for titanium (in which case the oxide dissolves into the metal, rather than the electrolyte) and fluoride ions for silicon. On the other hand, unusual conditions may result in passivation of materials that are normally unprotected, as the alkaline environment of concrete does for steel rebar. Exposure to a liquid metal such as mercury or hot solder can often circumvent passivation mechanisms.

## Corrosion in Passivated Materials

Passivation is extremely useful in mitigating corrosion damage, however even a high-quality alloy will corrode if its ability to form a passivating film is hindered. Proper selection of the right grade of material for the specific environment is important for the long-lasting performance of this group of materials. If breakdown occurs in the passive film due to chemical or mechanical factors, the resulting major modes of corrosion may include pitting corrosion, crevice corrosion and stress corrosion cracking.

## Pitting Corrosion

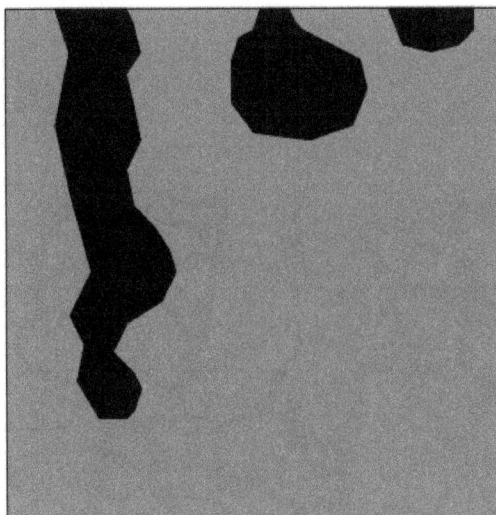

The scheme of pitting corrosion

Certain conditions, such as low concentrations of oxygen or high concentrations of species such as chloride which complete as anions, can interfere with a given alloy's ability to re-form a passivating film. In the worst case, almost all of the surface will remain protected, but tiny local fluctuations will degrade the oxide film in a few critical points. Corrosion at these points will be greatly amplified, and can cause *corrosion pits* of several types, depending upon conditions. While the corrosion pits only nucleate under fairly extreme circumstances, they can continue to grow even when conditions return to normal, since the interior of a pit is naturally deprived of oxygen and locally the pH decreases to very low values and the corrosion rate increases due to an autocatalytic process. In extreme cases, the sharp tips of extremely long and narrow corrosion pits can cause stress concentration to the point that otherwise tough alloys can shatter; a thin film pierced by an invisibly small hole can hide a thumb sized pit from view. These problems are especially dangerous because they are difficult to detect before a part or structure fails. Pitting remains among the most common and damaging forms of corrosion in passivated alloys, but it can be prevented by control of the alloy's environment.

Pitting results when a small hole, or cavity, forms in the metal, usually as a result of de-passivation of a small area. This area becomes anodic, while part of the remaining metal becomes cathodic, producing a localized galvanic reaction. The deterioration of this small area penetrates the metal and can lead to failure. This form of corrosion is often difficult to detect due to the fact that it is usually relatively small and may be covered and hidden by corrosion-produced compounds.

## Weld Decay and Knifeline Attack

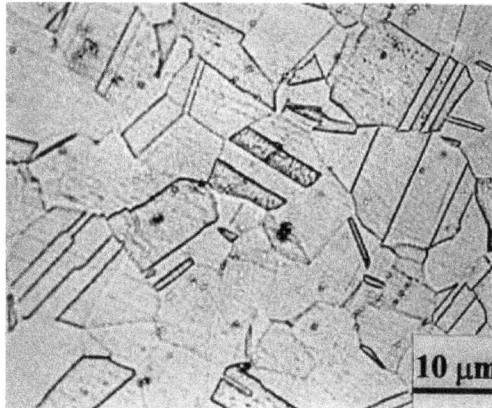

Normal microstructure

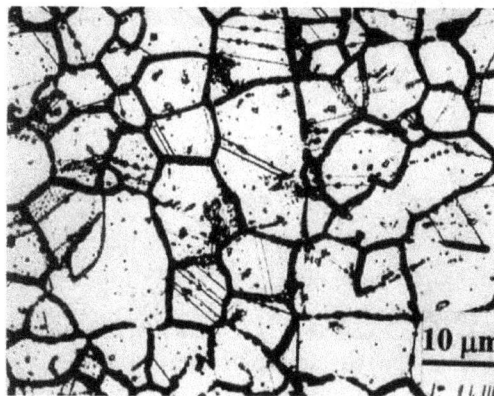

Sensitized microstructure

Stainless steel can pose special corrosion challenges, since its passivating behavior relies on the presence of a major alloying component (chromium, at least 11.5%). Because of the elevated temperatures of welding and heat treatment, chromium carbides can form in the grain boundaries of stainless alloys. This chemical reaction robs the material of chromium in the zone near the grain boundary, making those areas much less resistant to corrosion. This creates a galvanic couple with the well-protected alloy nearby, which leads to *weld decay* (corrosion of the grain boundaries in the heat affected zones) in highly corrosive environments.

A stainless steel is said to be sensitized if chromium carbides are formed in the microstructure. A typical microstructure of a normalized type 304 stainless steel shows no signs of sensitization while a heavily sensitized steel shows the presence of grain boundary precipitates. The dark lines in the sensitized microstructure are networks of chromium carbides formed along the grain boundaries.

Special alloys, either with low carbon content or with added carbon "getters" such as titanium and niobium (in types 321 and 347, respectively), can prevent this effect, but the latter require special heat treatment after welding to prevent the similar phenomenon of *knifeline attack*. As its name implies, corrosion is limited to a very narrow zone adjacent to the weld, often only a few micrometers across, making it even less noticeable.

## Crevice Corrosion

Crevice corrosion is a localized form of corrosion occurring in confined spaces (crevices), to which the access of the working fluid from the environment is limited. Formation of a differential aeration cell leads to corrosion inside the crevices. Examples of crevices are gaps and contact areas between parts, under gaskets or seals, inside cracks and seams, spaces filled with deposits and under sludge piles.

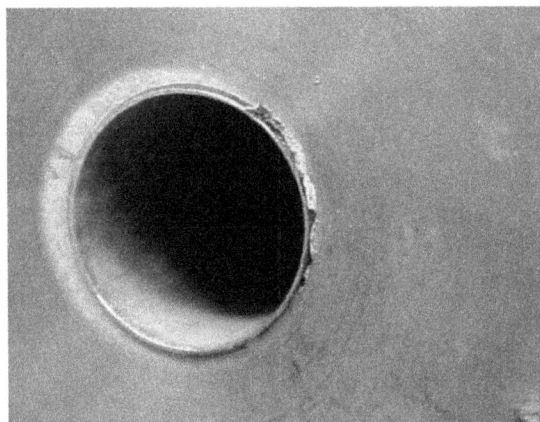

Corrosion in the crevice between the tube and tube sheet (both made of type 316 stainless steel) of a heat exchanger in a seawater desalination plant.

Crevice corrosion is influenced by the crevice type (metal-metal, metal-nonmetal), crevice geometry (size, surface finish), and metallurgical and environmental factors. The susceptibility to crevice corrosion can be evaluated with ASTM standard procedures. A critical crevice corrosion temperature is commonly used to rank a material's resistance to crevice corrosion.

## Microbial Corrosion

Microbial corrosion, or commonly known as microbiologically influenced corrosion (MIC), is a corrosion caused or promoted by microorganisms, usually chemoautotrophs. It can apply to both metallic and non-metallic materials, in the presence or absence of oxygen. Sulfate-reducing bacteria are active in the absence of oxygen (anaerobic); they produce hydrogen sulfide, causing sulfide stress cracking. In the presence of oxygen (aerobic), some bacteria may directly oxidize iron to iron oxides and hydroxides, other bacteria oxidize sulfur and produce sulfuric acid causing biogenic sulfide corrosion. Concentration cells can form in the deposits of corrosion products, leading to localized corrosion.

Accelerated low-water corrosion (ALWC) is a particularly aggressive form of MIC that affects steel piles in seawater near the low water tide mark. It is characterized by an orange sludge, which smells of hydrogen sulfide when treated with acid. Corrosion rates can be very high and design corrosion allowances can soon be exceeded leading to premature failure of the steel pile. Piles that have been coated and have cathodic protection installed at the time of construction are not susceptible to ALWC. For unprotected piles, sacrificial anodes can be installed local to the affected areas to inhibit the corrosion or a complete retrofitted sacrificial anode system can be installed. Affected areas can also be treated electrochemically by using an electrode to first produce chlorine to kill the bacteria, and then to produce a calcareous deposit, which will help shield the metal from further attack.

## High-temperature Corrosion

High-temperature corrosion is chemical deterioration of a material (typically a metal) as a result of heating. This non-galvanic form of corrosion can occur when a metal is subjected to a hot atmosphere containing oxygen, sulfur, or other compounds capable of oxidizing (or assisting the oxidation of) the material concerned. For example, materials used in aerospace, power generation and even in car engines have to resist sustained periods at high temperature in which they may be exposed to an atmosphere containing potentially highly corrosive products of combustion.

The products of high-temperature corrosion can potentially be turned to the advantage of the engineer. The formation of oxides on stainless steels, for example, can provide a protective layer preventing further atmospheric attack, allowing for a material to be used for sustained periods at both room and high temperatures in hostile conditions. Such high-temperature corrosion products, in the form of compacted oxide layer glazes, prevent or reduce wear during high-temperature sliding contact of metallic (or metallic and ceramic) surfaces.

## Metal Dusting

Metal dusting is a catastrophic form of corrosion that occurs when susceptible materials are exposed to environments with high carbon activities, such as synthesis gas and other high-CO environments. The corrosion manifests itself as a break-up of bulk metal to metal powder. The suspected mechanism is firstly the deposition of a graphite layer on the surface of the metal, usually from carbon monoxide (CO) in the vapor phase. This graphite layer is then thought to form metastable $M_3C$ species (where M is the metal), which migrate away from the metal surface. However, in some regimes no $M_3C$ species is observed indicating a direct transfer of metal atoms into the graphite layer.

## Protection from Corrosion

The US Army shrink wraps equipment such as helicopters to protect them from corrosion and thus save millions of dollars.

## Surface Treatments

## Applied Coatings

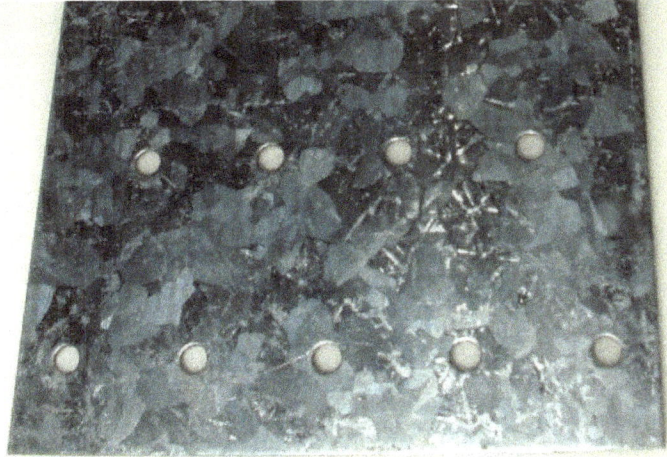

Galvanized surface

Plating, painting, and the application of enamel are the most common anti-corrosion treatments. They work by providing a barrier of corrosion-resistant material between the damaging environment and the structural material. Aside from cosmetic and manufacturing issues, there may be tradeoffs in mechanical flexibility versus resistance to abrasion and high temperature. Platings usually fail only in small sections, but if the plating is more noble than the substrate (for example, chromium on steel), a galvanic couple will cause any exposed area to corrode much more rapidly than an unplated surface would. For this reason, it is often wise to plate with active metal such as zinc or cadmium.

Painting either by roller or brush is more desirable for tight spaces; spray would be better for larger coating areas such as steel decks and waterfront applications. Flexible polyurethane coatings, like Durabak-M26 for example, can provide an anti-corrosive seal with a highly durable slip resistant membrane. Painted coatings are relatively easy to apply and have fast drying times although temperature and humidity may cause dry times to vary.

## Reactive Coatings

If the environment is controlled (especially in recirculating systems), corrosion inhibitors can often be added to it. These chemicals form an electrically insulating or chemically impermeable coating on exposed metal surfaces, to suppress electrochemical reactions. Such methods make the system less sensitive to scratches or defects in the coating, since extra inhibitors can be made available wherever metal becomes exposed. Chemicals that inhibit corrosion include some of the salts in hard water (Roman water systems are famous for their mineral deposits), chromates, phosphates, polyaniline, other conducting polymers and a wide range of specially-designed chemicals that resemble surfactants (i.e. long-chain organic molecules with ionic end groups).

## Anodization

Aluminium alloys often undergo a surface treatment. Electrochemical conditions in the bath are carefully adjusted so that uniform pores, several nanometers wide, appear in the metal's oxide film. These pores allow the oxide to grow much thicker than passivating conditions would allow. At the end of the treatment, the pores are allowed to seal, forming a harder-than-usual surface layer. If this coating is scratched, normal passivation processes take over to protect the damaged area.

This climbing descender is anodized with a yellow finish.

Anodizing is very resilient to weathering and corrosion, so it is commonly used for building facades and other areas where the surface will come into regular contact with the elements. While being resilient, it must be cleaned frequently. If left without cleaning, panel edge staining will naturally occur.

## Biofilm Coatings

A new form of protection has been developed by applying certain species of bacterial films to the surface of metals in highly corrosive environments. This process increases the corrosion resistance substantially. Alternatively, antimicrobial-producing biofilms can be used to inhibit mild steel corrosion from sulfate-reducing bacteria.

## Controlled Permeability Formwork

Controlled permeability formwork (CPF) is a method of preventing the corrosion of reinforcement by naturally enhancing the durability of the cover during concrete placement. CPF has been used in environments to combat the effects of carbonation, chlorides, frost and abrasion.

## Cathodic Protection

Cathodic protection (CP) is a technique to control the corrosion of a metal surface by making that surface the cathode of an electrochemical cell. Cathodic protection systems are most commonly used to protect steel, and pipelines and tanks; steel pier piles, ships, and offshore oil platforms.

## Sacrificial Anode Protection

For effective CP, the potential of the steel surface is polarized (pushed) more negative until the metal surface has a uniform potential. With a uniform potential, the driving force for the corrosion reaction is halted. For galvanic CP systems, the anode material corrodes under the influence of the steel, and eventually it must be replaced. The polarization is caused by the current flow from the anode to the cathode, driven by the difference in electrode potential between the anode and the cathode.

Sacrificial anode on the hull of a ship.

## Impressed Current Cathodic Protection

For larger structures, galvanic anodes cannot economically deliver enough current to provide complete protection. Impressed current cathodic protection (ICCP) systems use anodes connected to a DC power source (such as a cathodic protection rectifier). Anodes for ICCP systems are tubular and solid rod shapes of various specialized materials. These include high silicon cast iron, graphite, mixed metal oxide or platinum coated titanium or niobium coated rod and wires.

## Anodic Protection

Anodic protection impresses anodic current on the structure to be protected (opposite to the cathodic protection). It is appropriate for metals that exhibit passivity (e.g. stainless steel) and suitably small passive current over a wide range of potentials. It is used in aggressive environments, such as solutions of sulfuric acid.

## Rate of Corrosion

A simple test for measuring corrosion is the weight loss method. The method involves exposing a clean weighed piece of the metal or alloy to the corrosive environment for a specified time followed by cleaning to remove corrosion products and weighing the piece to determine the loss of weight. The rate of corrosion (R) is calculated as

$$R = \frac{kW}{\rho At}$$

where $k$ is a constant, $W$ is the weight loss of the metal in time $t$, $A$ is the surface area of the metal exposed, and $\rho$ is the density of the metal (in $g/cm^3$).

Other common expressions for the corrosion rate is penetration depth and change of mechanical properties.

## Economic Impact

The collapsed Silver Bridge, as seen from the Ohio side

In 2002, the US Federal Highway Administration released a study titled "Corrosion Costs and Preventive Strategies in the United States" on the direct costs associated with metallic corrosion in the US industry. In 1998, the total annual direct cost of corrosion in the U.S. was ca. $276 billion (ca. 3.2% of the US gross domestic product). Broken down into five specific industries, the economic losses are $22.6 billion in infrastructure; $17.6 billion in production and manufacturing; $29.7 billion in transportation; $20.1 billion in government; and $47.9 billion in utilities.

Rust is one of the most common causes of bridge accidents. As rust has a much higher volume than the originating mass of iron, its build-up can also cause failure by forcing apart adjacent parts. It was the cause of the collapse of the Mianus river bridge in 1983, when the bearings rusted internally and pushed one corner of the road slab off its support. Three drivers on the roadway at the time died as the slab fell into the river below. The following NTSB investigation showed that a drain in the road had been blocked for road re-surfacing, and had not been unblocked; as a result, runoff water penetrated the support hangers. Rust was also an important factor in the Silver Bridge disaster of 1967 in West Virginia, when a steel suspension bridge collapsed within a minute, killing 46 drivers and passengers on the bridge at the time.

Similarly, corrosion of concrete-covered steel and iron can cause the concrete to spall, creating severe structural problems. It is one of the most common failure modes of reinforced concrete bridges. Measuring instruments based on the half-cell potential can detect the potential corrosion spots before total failure of the concrete structure is reached.

Until 20–30 years ago, galvanized steel pipe was used extensively in the potable water systems for single and multi-family residents as well as commercial and public construction. Today, these systems have long ago consumed the protective zinc and are corroding internally resulting in poor water quality and pipe failures. The economic impact on homeowners, condo dwellers, and the public infrastructure is estimated at 22 billion dollars as the insurance industry braces for a wave of claims due to pipe failures.

## Corrosion in Nonmetals

Most ceramic materials are almost entirely immune to corrosion. The strong chemical bonds that hold them together leave very little free chemical energy in the structure; they can be thought of as already corroded. When corrosion does occur, it is almost always a simple dissolution of the material or chemical reaction, rather than an electrochemical process. A common example of corrosion protection in ceramics is the lime added to soda-lime glass to reduce its solubility in water; though it is not nearly as soluble as pure sodium silicate, normal glass does form sub-microscopic flaws when exposed to moisture. Due to its brittleness, such flaws cause a dramatic reduction in the strength of a glass object during its first few hours at room temperature.

## Corrosion of Polymers

Ozone cracking in natural rubber tubing

Polymer degradation involves several complex and often poorly understood physiochemical processes. These are strikingly different from the other processes discussed here, and so the term "corrosion" is only applied to them in a loose sense of the word. Because of their large molecular weight, very little entropy can be gained by mixing a given mass of polymer with another substance, making them generally quite difficult to dissolve. While dissolution is a problem in some polymer applications, it is relatively simple to design against.

A more common and related problem is *swelling*, where small molecules infiltrate the structure, reducing strength and stiffness and causing a volume change. Conversely, many polymers (notably flexible vinyl) are intentionally swelled with plasticizers, which can be leached out of the structure, causing brittleness or other undesirable changes.

The most common form of degradation, however, is a decrease in polymer chain length. Mechanisms which break polymer chains are familiar to biologists because of their effect on DNA: ionizing radiation (most commonly ultraviolet light), free radicals, and oxidizers such as oxygen, ozone, and chlorine. Ozone cracking is a well-known problem affecting natural rubber for example. Additives can slow these process very effectively, and can be as simple as a UV-absorbing pigment (e.g. titanium dioxide or carbon black). Plastic shopping bags often do not include these additives so that they break down more easily as ultrafine particles of litter.

## Corrosion of Glasses

Glass is characterized by a high degree of corrosion-resistance. Because of its high water-resistance it is often used as primary packaging material in the pharma industry since most medicines are preserved in a watery solution. Besides its water-resistance, glass is also very robust when being exposed to chemically aggressive liquids or gases. While other materials like metal or plastics quickly reach their limits, special glass-types can easily hold up.

Glass corrosion

Glass disease is the corrosion of silicate glasses in aqueous solutions. It is governed by two mechanisms: diffusion-controlled leaching (ion exchange) and hydrolytic dissolution of the glass network. Both mechanisms strongly depend on the pH of contacting solution: the rate of ion exchange decreases with pH as $10^{-0.5pH}$ whereas the rate of hydrolytic dissolution increases with pH as $10^{0.5pH}$.

Mathematically, corrosion rates of glasses are characterized by normalized corrosion rates of elements $NR_i$ (g/cm²·d) which are determined as the ratio of total amount of released species into the water $M_i$ (g) to the water-contacting surface area S (cm²), time of contact t (days) and weight fraction content of the element in the glass $f_i$:

$$NR_i = \frac{M_i}{Sf_i t}.$$

The overall corrosion rate is a sum of contributions from both mechanisms (leaching + dissolution) $NR_i=Nrx_i+NRh$. Diffusion-controlled leaching (ion exchange) is characteristic of the initial phase of corrosion and involves replacement of alkali ions in the glass by a hydronium ($H_3O^+$) ion from the solution. It causes an ion-selective depletion of near surface layers of glasses and gives an inverse square root dependence of corrosion rate with exposure time. The diffusion-controlled normalized leaching rate of cations from glasses (g/cm²·d) is given by:

$$NRx_i = 2\rho\sqrt{\frac{D_i}{\pi t}},$$

where t is time, $D_i$ is the i-th cation effective diffusion coefficient (cm²/d), which depends on pH of contacting water as $D_i = D_{io} \cdot 10^{-pH}$, and $\rho$ is the density of the glass (g/cm³).

Glass network dissolution is characteristic of the later phases of corrosion and causes a congruent release of ions into the water solution at a time-independent rate in dilute solutions (g/cm²·d):

$$NRh = \rho r_h,$$

where $r_h$ is the stationary hydrolysis (dissolution) rate of the glass (cm/d). In closed systems the consumption of protons from the aqueous phase increases the pH and causes a fast transition to hydrolysis. However, a further saturation of solution with silica impedes hydrolysis and causes the glass to return to an ion-exchange, e.g. diffusion-controlled regime of corrosion.

In typical natural conditions normalized corrosion rates of silicate glasses are very low and are of the order of $10^{-7}$–$10^{-5}$ g/(cm²·d). The very high durability of silicate glasses in water makes them suitable for hazardous and nuclear waste immobilisation.

## Glass Corrosion Tests

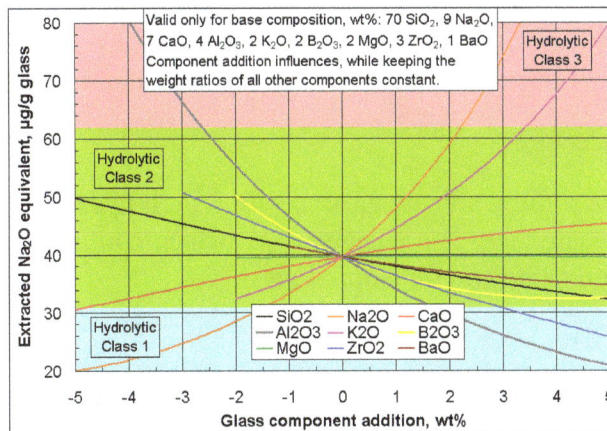

Effect of addition of a certain glass component on the chemical durability against water corrosion of a specific base glass (corrosion test ISO 719).

There exist numerous standardized procedures for measuring the corrosion (also called chemical durability) of glasses in neutral, basic, and acidic environments, under simulated environmental conditions, in simulated body fluid, at high temperature and pressure, and under other conditions.

The standard procedure ISO 719 describes a test of the extraction of water-soluble basic compounds under neutral conditions: 2 g of glass, particle size 300–500 µm, is kept for 60 min in 50 ml de-ionized water of grade 2 at 98 °C; 25 ml of the obtained solution is titrated against 0.01 mol/l HCl solution. The volume of HCl required for neutralization is classified according to the table below.

| Amount of 0.01M HCl needed to neutralize extracted basic oxides, ml | Extracted $Na_2O$ equivalent, µg | Hydrolytic class |
|---|---|---|
| < 0.1 | < 31 | 1 |
| 0.1-0.2 | 31-62 | 2 |
| 0.2-0.85 | 62-264 | 3 |
| 0.85-2.0 | 264-620 | 4 |
| 2.0-3.5 | 620-1085 | 5 |
| > 3.5 | > 1085 | > 5 |

The standardized test ISO 719 is not suitable for glasses with poor or not extractable alkaline components, but which are still attacked by water, e.g. quartz glass, $B_2O_3$ glass or $P_2O_5$ glass.

Usual glasses are differentiated into the following classes:

Hydrolytic class 1 (Type I):

This class, which is also called neutral glass, includes borosilicate glasses (e.g. Duran, Pyrex, Fiolax).

Glass of this class contains essential quantities of boron oxides, aluminium oxides and alkaline earth oxides. Through its composition neutral glass has a high resistance against temperature shocks and the highest hydrolytic resistance. Against acid and neutral solutions it shows high chemical resistance, because of its poor alkali content against alkaline solutions.

Hydrolytic class 2 (Type II):

This class usually contains sodium silicate glasses with a high hydrolytic resistance through surface finishing. Sodium silicate glass is a silicate glass, which contains alkali- and alkaline earth oxide and primarily sodium oxide and Calcium oxide.

Hydrolytic class 3 (Type III):

Glass of the 3rd hydrolytic class usually contains sodium silicate glasses and has a mean hydrolytic resistance, which is two times poorer than of type 1 glasses.

Acid class DIN 12116 and alkali class DIN 52322 (ISO 695) are to be distinguished from the hydrolytic class DIN 12111 (ISO 719).

# Anaerobic Corrosion

Hydrogen corrosion is a form of metal corrosion occurring in the presence of anoxic water. Hydrogen corrosion involves a redox reaction that reduces hydrogen ions, forming molecular hydrogen.

Metals enter aqueous solution and are oxidized.

*Oxidation reaction (pH independent):*

$$Fe \rightarrow Fe^{2+} + 2e^-$$

*Reduction reaction in acid solution:*

$$2H^+ + 2e^- \rightarrow H_2$$

In an acidic solution, the water molecules are protonated and the hydronium ions ($H_3O^+$) are directly reduced into $H_2$.

*Reduction reaction in neutral or slightly alkaline solution:*

$$2H_2O + 2e^- \rightarrow H_2 + 2OH^-$$

In a neutral or slightly alkaline solution, the protons of water are reduced into molecular hydrogen giving rise to the production of hydroxide ions responsible of the precipitation of the slightly soluble ferrous hydroxide ($Fe(OH)_2$).

This finally leads to the global reaction of the anaerobic corrosion of iron in water:

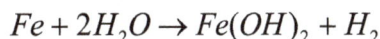

$$Fe + 2H_2O \rightarrow Fe(OH)_2 + H_2$$

## Transformation of Ferrous Hydroxide into Magnetite

Under anaerobic conditions, the ferrous hydroxide ($Fe(OH)_2$) can be oxidized by the protons of water to form magnetite and molecular hydrogen. This process is described by the Schikorr reaction:

$$3\ Fe(OH)_2 \rightarrow Fe_3O_4 + H_2 + 2\ H_2O$$

*ferrous hydroxide → magnetite + hydrogen + water*

The well crystallized magnetite ($Fe_3O_4$) is thermodynamically more stable than the ferrous hydroxide ($Fe(OH)_2$).

This process also occurs during the anaerobic corrosion of iron and steel in oxygen-free groundwater and in reducing soils below the water table.

# Cyclic Corrosion Testing

Cyclic Corrosion Testing (CCT) has evolved in recent years, largely within the automotive industry, as a way of accelerating real-world corrosion failures, under laboratory controlled conditions. As the name implies, the test comprises different climates which are cycled automatically so the samples under test undergo the same sort of changing environment that would be encountered in the natural world. The intention being to bring about the type of failure that might occur naturally, but more quickly i.e. accelerated. By doing this manufacturers and suppliers can predict, more accurately, the service life expectancy of their products.

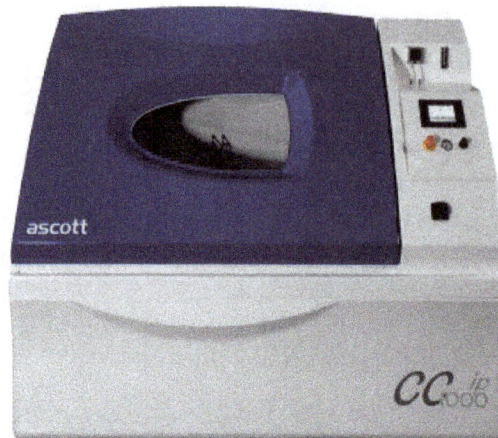

Example of a Cyclic corrosion test chamber.

Until the development of Cyclic Corrosion Testing, the traditional Salt spray test was virtually all that manufacturers could use for this purpose. However, this test was never intended for this purpose. Because the test conditions specified for salt spray testing are not typical of a naturally occurring environment, this type of test cannot be used as a reliable means of predicting the 'real world' service life expectancy for the samples under test. The sole purpose of the salt spray test is to compare and contrast results with previous experience to perform a quality audit. So, for example, a spray test can be used to 'police' a production process and forewarn of potential manufacturing problems or defects, which might affect corrosion resistance. .

To recreate these different environments within an environmental chamber requires much more flexible testing procedures than are available in a standard salt spray chamber.

The lack of correlation between results obtained from traditional salt spray testing and the 'real world' atmospheric corrosion of vehicles, left the automotive industry without a reliable test method for predicting the service life expectancy of their products. This was and remains of particular concern in an industry where anti-corrosion warranties have been gradually increasing and now run to several years for new vehicles.

With ever increasing consumer pressure for improved vehicle corrosion resistance and a few 'high profile' corrosion failures amongst some vehicle manufactures – with disastrous commercial consequences, the automotive industry recognized the need for a different type of corrosion test.

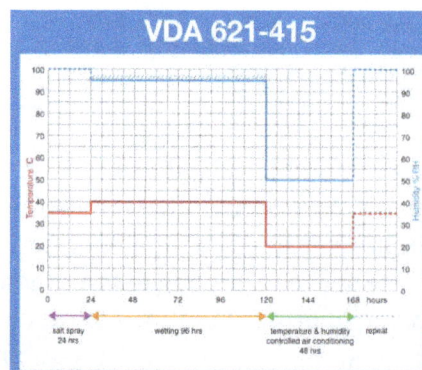

Graph showing the temperature & humidity steps required during cyclic corrosion test VDA 621-415

Such a test would need to simulate the types of conditions a vehicle might encounter naturally, but recreate and accelerate these conditions, with good repeatability, within the convenience of the laboratory. CCT is effective for evaluating a variety of corrosion types, including galvanic corrosion and crevice corrosion.

## Test Stages

Taking results gathered largely from 'real world' exposure sites, automotive companies, led originally by the Japanese automobile industry, developed their own Cyclic Corrosion Tests. These have evolved in different ways for different vehicle manufacturers, and such tests still remain largely industry specific, with no truly international CCT standard. However, they all generally require most of the following conditions to be created, in a repeating sequence or 'cycle', though not necessarily in the following order:

• A salt spray 'pollution' phase. This may be similar to the traditional salt spray test although in some cases direct impingement by the salt solution on the test specimens, or even complete immersion in salt water, is required. However, this 'pollution' phase is generally shorter in duration than a traditional salt spray test.

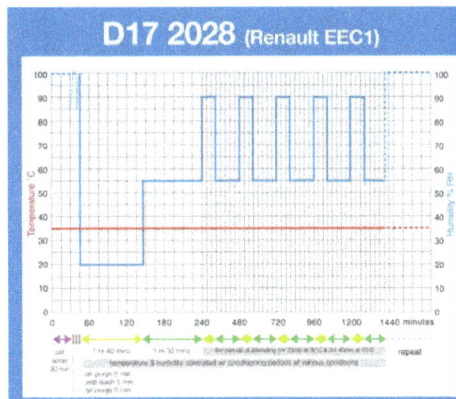

Graph showing the temperature & humidity steps required during cyclic corrosion test D17 2028 ECC1

• An air drying phase. Depending on the test, this may be conducted at ambient temperature, or at an elevated temperature, with or without control over the relative humidity and usually by introducing a continuous supply of relatively fresh air around the test samples at the same time. It is generally required that the samples under test should be visibly 'dry' at the end of this test phase.

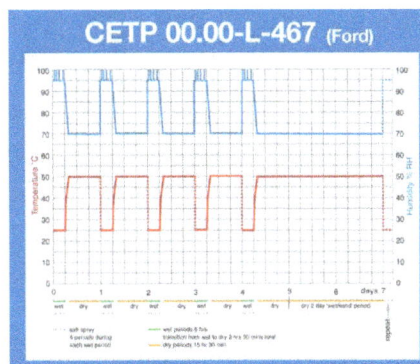

Graph showing the temperature & humidity steps required during cyclic corrosion test CETP 00.00-L-467

• A condensation humidity 'wetting' phase. This is usually conducted at an elevated temperature and generally a high humidity of 95-100%RH. The purpose of this phase is to promote the formation of condensation on the surfaces of the samples under test.

• A controlled humidity/humidity cycling phase. This requires the tests samples to be exposed to a controlled temperature and controlled humidity climate, which can either be constant or cycling between different levels. When cycling between different levels, the rate of change may also be specified.

The above list is not exhaustive, since some automotive companies may also require other climates to be created in sequence as well, for example; sub-zero refrigeration, but it does list the most common requirements.

## Tests Standards

The below list is not exhaustive, but here are some examples of popular cyclic corrosion test standards,

- ACT 1 (Volvo)

- ACT 2 (Volvo)

- CETP 00.00-L-467 (Ford)

- D17 2028 (Renault)

- JASO M 609

- SAE J 2334

- VDA 621-415

## References

- JE Breakell, M Siegwart, K Foster, D Marshall, M Hodgson, R Cottis, S Lyon. Management of Accelerated Low Water Corrosion in Steel Maritime Structures, Volume 634 of CIRIA Series, 2005, ISBN 0-86017-634-7

- [Fundamentals of corrosion - Mechanisms, Causes and Preventative Methods]. Philip A. Schweitzer, Taylor and Francis Group, LLC (2010) ISBN 978-1-4200-6770-5, p. 25.

- M.I. Ojovan, W.E. Lee. New Developments in Glassy Nuclear Wasteforms. Nova Science Publishers, New York (2007) ISBN 1600217834 pp.

- Corrosion of Glass, Ceramics and Ceramic Superconductors. D.E. Clark, B.K. Zoitos (eds.), William Andrew Publishing/Noyes (1992) ISBN 081551283X.

- Daniel Robles. "Potable Water Pipe Condition Assessment For a High Rise Condominium in The Pacific Northwest". GSG Group, Inc. Dan Robles, PE. Retrieved 10 December 2012.

- International Organization for Standardization, Procedure 719 (1985). Iso.org (2011-01-21). Retrieved on 2012-07-15.

# Metalworking and its Processes

The process by which various parts, machinery, jewelry, engines and assemblies are created from metals is known as metalworking. The chapter details the processes of casting, rolling, hot working, sheet metal etc. The reader is introduced to the workings of a foundry and the processes involved in cold formed steel. This chapter is a compilation of the various branches of metalworking that form an integral part of the broader subject matter.

## Metalworking

Metalworking is the process of working with metals to create individual parts, assemblies, or large-scale structures. The term covers a wide range of work from large ships and bridges to precise engine parts and delicate jewelry. It therefore includes a correspondingly wide range of skills, processes, and tools.

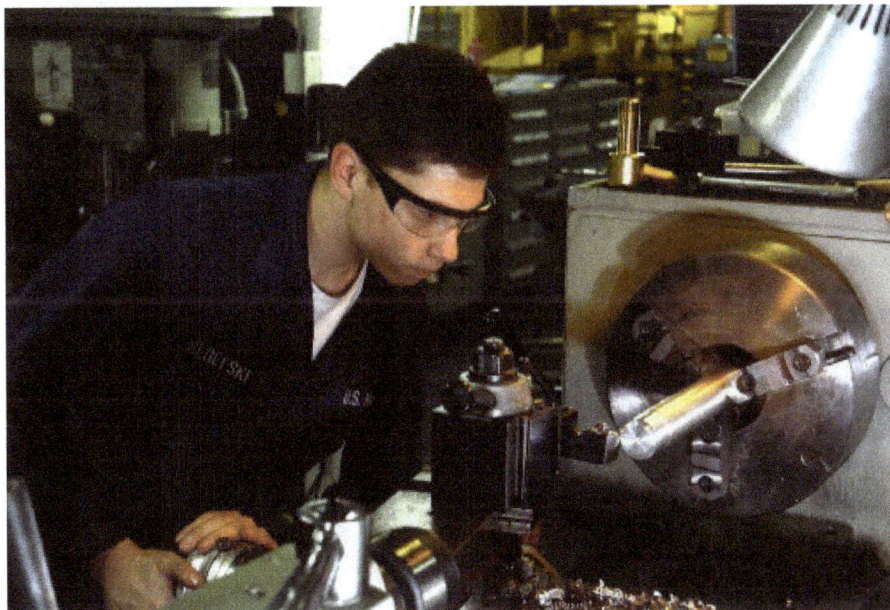

Turning a bar of metal on a lathe

Metalworking is a science, art, hobby, industry and trade. Its historical roots span cultures, civilizations, and millennia. Metalworking has evolved from the discovery of smelting various ores, producing malleable and ductile metal useful for tools and adornments. Modern metalworking processes, though diverse and specialized, can be categorized as forming, cutting, or joining processes. Today's machine shop includes a number of machine tools capable of creating a precise, useful workpiece.

## Prehistory

The oldest archaeological evidence of copper mining and working was the discovery of a copper pendant in northern Iraq from 8,700 BCE. The earliest substantiated and dated evidence of metalworking in the Americas was the processing of copper in Wisconsin, near Lake Michigan. Copper was hammered until brittle then heated so it could be worked some more. This technology is dated to about 4000–5000 BCE. The oldest gold artifacts in the world come from the Bulgarian Varna Necropolis and date from 4450 BCE.

Not all metal required fire to obtain it or work it. Isaac Asimov speculated that gold was the "first metal." His reasoning is that by its chemistry it is found in nature as nuggets of pure gold. In other words, gold, as rare as it is, is sometimes found in nature as the metal that it is. There are a few other metals that sometimes occur natively, and as a result of meteors. Almost all other metals are found in ores, a mineral-bearing rock, that require heat or some other process to liberate the metal. Another feature of gold is that it is workable as it is found, meaning that no technology beyond a stone hammer and anvil to work the metal is needed. This is a result of gold's properties of malleability and ductility. The earliest tools were stone, bone, wood, and sinew, all of which sufficed to work gold.

At some unknown point the connection between heat and the liberation of metals from rock became clear, rocks rich in copper, tin, and lead came into demand. These ores were mined wherever they were recognized. Remnants of such ancient mines have been found all over Southwestern Asia. Metalworking was being carried out by the South Asian inhabitants of Mehrgarh between 7000–3300 BCE. The end of the beginning of metalworking occurs sometime around 6000 BCE when copper smelting became common in Southwestern Asia.

Ancient civilisations knew of seven metals. Here they are arranged in order of their oxidation potential (in volts):

- Iron +0.44 V,
- Tin +0.14 V
- Lead +0.13 V
- Copper –0.34 V
- Mercury –0.79 V
- Silver –0.80 V
- Gold –1.50 V.

The oxidation potential is important because it is one indicator of how tightly bound to the ore the metal is likely to be. As can be seen, iron is significantly higher than the other six metals while gold is dramatically lower than the six above it. Gold's low oxidation is one of the main reasons that gold is found in nuggets. These nuggets are relatively pure gold and are workable as they are found.

Copper ore, being relatively abundant, and tin ore became the next important players in the story of metalworking. Using heat to smelt copper from ore, a great deal of copper was produced. It was used for both jewelry and simple tools. However, copper by itself was too soft for tools requiring edges and stiffness. At some point tin was added into the molten copper and bronze was born. Bronze is an alloy of copper and tin. Bronze was an important advance because it had the edge-durability and stiffness that pure copper lacked. Until the advent of iron, bronze was the most advanced metal for tools and weapons in common use.

Outside Southwestern Asia, these same advances and materials were being discovered and used around the world. China and Great Britain jumped into the use of bronze with little time being devoted to copper. Japan began the use of bronze and iron almost simultaneously. In the Americas things were different. Although the peoples of the Americas knew of metals, it wasn't until the European colonisation that metalworking for tools and weapons became common. Jewelry and art were the principal uses of metals in the Americas prior to European influence.

Around 2700 BCE, production of bronze was common in locales where the necessary materials could be assembled for smelting, heating, and working the metal. Iron was beginning to be smelted and began its emergence as an important metal for tools and weapons. The Iron Age was dawning.

## History

A turret lathe operator machining parts for transport planes at the Consolidated Aircraft Corporation plant, Fort Worth, Texas, USA in the 1940s

By the historical periods of the Pharaohs in Egypt, the Vedic Kings in India, the Tribes of Israel, and the Maya civilization in North America, among other ancient populations, precious metals began to have value attached to them. In some cases rules for ownership, distribution, and trade were created, enforced, and agreed upon by the respective peoples. By the above periods metalworkers were very skilled at creating objects of adornment, religious artifacts, and trade instruments of precious metals (non-ferrous), as well as weaponry usually of ferrous metals and/or alloys. These skills were finely honed and well executed. The techniques were practiced by artisans, blacksmiths, atharvavedic practitioners, alchemists, and other categories of metalworkers around the globe. For example, the ancient technique of granulation is found around the world in numerous ancient cultures before the historic record shows people traveled to far regions to share this process that is still being used by metalsmiths today.

As time progressed metal objects became more common, and ever more complex. The need to further acquire and work metals grew in importance. Skills related to extracting metal ores from the earth began to evolve, and metalsmiths became more knowledgeable. Metalsmiths became important members of society. Fates and economies of entire civilizations were greatly affected by the availability of metals and metalsmiths. The metalworker depends on the extraction of precious metals to make jewelry, build more efficient electronics, and for industrial and technological applications from construction to shipping containers to rail, and air transport. Without metals, goods and services would cease to move around the globe on the scale we know today.

## General Metalworking Processes

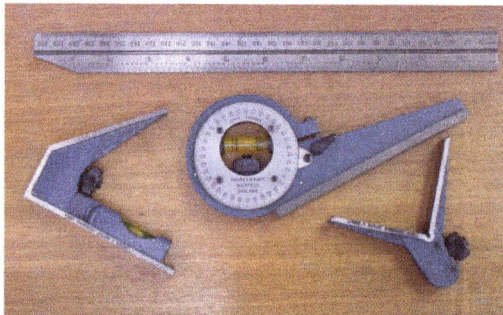

A combination square used for transferring designs.

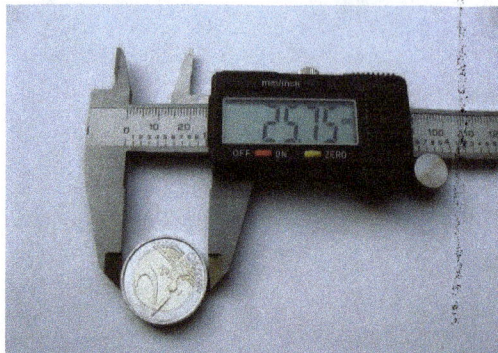

A caliper is used to precisely measure a short length.

Metalworking generally is divided into the following categories, *forming*, *cutting*, and, *joining*. Each of these categories contain various processes.

Prior to most operations, the metal must be marked out and/or measured, depending on the desired finished product.

*Marking out* (also known as layout) is the process of transferring a design or pattern to a workpiece and is the first step in the handcraft of metalworking. It is performed in many industries or hobbies, although in industry, the repetition eliminates the need to mark out every individual piece. In the metal trades area, marking out consists of transferring the engineer's plan to the workpiece in preparation for the next step, machining or manufacture.

*Calipers* are hand tools designed to precisely measure the distance between two points. Most calipers have two sets of flat, parallel edges used for inner or outer diameter measurements. These calipers can be accurate to within one-thousandth of an inch (25.4 μm). Different types of calipers have different mechanisms for displaying the distance measured. Where larger objects need to be measured with less precision, a tape measure is often used.

| Compatibility chart of materials versus processes | | | | | | | | | | | |
|---|---|---|---|---|---|---|---|---|---|---|---|
| **Process** | **Material** | | | | | | | | | | |
| | **Iron** | **Steel** | **Aluminium** | **Copper** | **Magnesium** | **Nickel** | **Refractory metals** | **Titanium** | **Zinc** | **Brass** | **Bronze** |
| Sand casting | X | X | X | X | X | X | | | o | o | X |
| Permanent mold casting | X | o | X | o | X | o | | | o | o | X |
| Die casting | | | X | o | X | | | | X | | |
| Investment casting | | X | X | X | o | o | | | | o | X |
| Ablation casting | | X | X | X | o | o | | | | | |
| Closed-die forging | | X | o | o | o | o | o | o | | | |
| Extrusion | | o | X | X | X | o | o | o | | | |
| Cold heading | | X | X | X | | o | | | | | |
| Stamping & deep drawing | | X | X | X | o | X | | o | o | | |
| Screw machine | o | X | X | X | o | X | o | o | o | X | X |
| Powder metallurgy | X | X | o | X | | o | X | o | | | |
| Key: **X** = Routinely performed, **o** = Performed with difficulty, caution, or some sacrifice, **blank** = Not recommended | | | | | | | | | | | |

## Casting

A sand casting mold

Casting achieves a specific form by pouring molten metal into a mold and allowing it to cool, with no mechanical force. Forms of casting include:

- Investment casting (called lost wax casting in art)

- Centrifugal casting

- Die casting

- Sand casting

- Shell casting

- Spin casting

## Forming Processes

These *forming* processes modify metal or workpiece by deforming the object, that is, without removing any material. Forming is done with a system of mechanical forces and, especially for bulk metal forming, with heat.

## Bulk Forming Processes

A red-hot metal workpiece is inserted into a forging press.

Plastic deformation involves using heat or pressure to make a workpiece more conductive to mechanical force. Historically, this and casting were done by blacksmiths, though today the process has been industrialized. In bulk metal forming, the workpiece is generally heated up.

- Cold sizing

- Extrusion

- Drawing

- Forging

- Powder metallurgy

- Friction drilling

- Rolling

## Sheet (and Tube) Forming Processes

These types of forming process involve the application of mechanical force at room temperature. However, some recent developments involve the heating of dies and/or parts.

- Bending

- Coining

- Decambering

- Deep drawing (DD)

- Flowforming

- Hydroforming (HF)

- Hot metal gas forming

- Hot press hardening

- Incremental forming (IF)

- Spinning, Shear forming or Flowforming

A metal spun brass vase

- Raising

- Roll forming

- Roll bending

- Repoussé and chasing

- Rubber pad forming

- Shearing

- Stamping

- Superplastic forming (SPF)

- Wheeling using an English wheel (wheeling machine)

## Cutting Processes

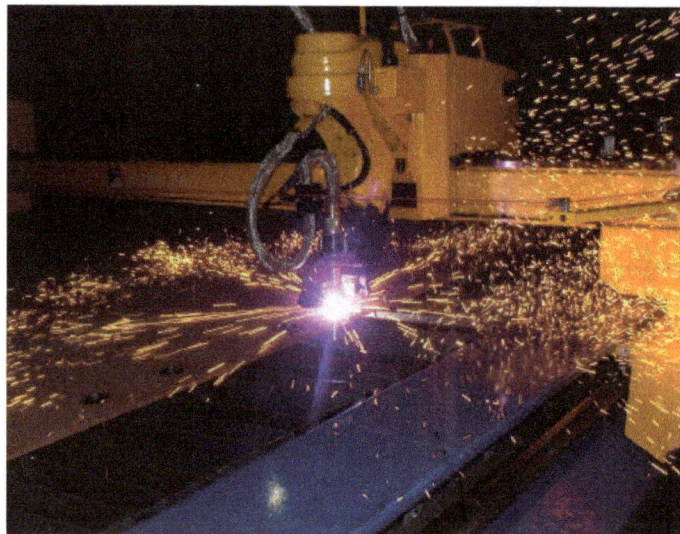

A CNC plasma cutting machining

*Cutting* is a collection of processes wherein material is brought to a specified geometry by removing excess material using various kinds of tooling to leave a finished part that meets specifications. The net result of cutting is two products, the waste or excess material, and the finished part. In woodworking, the waste would be sawdust and excess wood. In cutting metals the waste is chips or swarf and excess metal.

Cutting processes fall into one of three major categories:

- Chip producing processes most commonly known as machining

- Burning, a set of processes wherein the metal is cut by oxidizing a kerf to separate pieces of metal

- Miscellaneous specialty process, not falling easily into either of the above categories

Drilling a hole in a metal part is the most common example of a chip producing process. Using an oxy-fuel cutting torch to separate a plate of steel into smaller pieces is an example of burning. Chemical milling is an example of a specialty process that removes excess material by the use of etching chemicals and masking chemicals.

There are many technologies available to cut metal, including:

- Manual technologies: saw, chisel, shear or snips
- Machine technologies: turning, milling, drilling, grinding, sawing
- Welding/burning technologies: burning by laser, oxy-fuel burning, and plasma
- Erosion technologies: by water jet, electric discharge, or abrasive flow machining.
- Chemical technologies: Photochemical machining

Cutting fluid or coolant is used where there is significant friction and heat at the cutting interface between a cutter such as a drill or an end mill and the workpiece. Coolant is generally introduced by a spray across the face of the tool and workpiece to decrease friction and temperature at the cutting tool/workpiece interface to prevent excessive tool wear. In practice there are many methods of delivering coolant.

## Milling

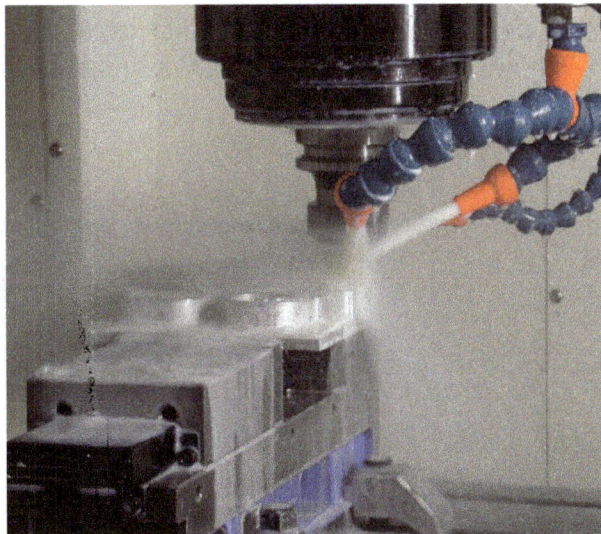

A milling machine in operation, including coolant hoses.

Milling is the complex shaping of metal or other materials by removing material to form the final shape. It is generally done on a milling machine, a power-driven machine that in its basic form consists of a milling cutter that rotates about the spindle axis (like a drill), and a worktable that can move in multiple directions (usually two dimensions [x and y axis] relative to the workpiece). The spindle usually moves in the z axis. It is possible to raise the table (where the workpiece rests). Milling machines may be operated manually or under computer numerical control (CNC), and can perform a vast number of complex operations, such as slot cutting, planing, drilling and threading, rabbeting, routing, etc. Two common types of mills are the horizontal mill and vertical mill.

The pieces produced are usually complex 3D objects that are converted into x, y, and z coordinates that are then fed into the CNC machine and allow it to complete the tasks required. The milling machine can produce most parts in 3D, but some require the objects to be rotated around the x, y, or z coordinate axis (depending on the need). Tolerances are usually in the thousandths of an inch (Unit known as Thou), depending on the specific machine.

In order to keep both the bit and material cool, a high temperature coolant is used. In most cases the coolant is sprayed from a hose directly onto the bit and material. This coolant can either be machine or user controlled, depending on the machine.

Materials that can be milled range from aluminum to stainless steel and almost everything in between. Each material requires a different speed on the milling tool and varies in the amount of material that can be removed in one pass of the tool. Harder materials are usually milled at slower speeds with small amounts of material removed. Softer materials vary, but usually are milled with a high bit speed.

The use of a milling machine adds costs that are factored into the manufacturing process. Each time the machine is used coolant is also used, which must be periodically added in order to prevent breaking bits. A milling bit must also be changed as needed in order to prevent damage to the material. Time is the biggest factor for costs. Complex parts can require hours to complete, while very simple parts take only minutes. This in turn varies the production time as well, as each part will require different amounts of time.

Safety is key with these machines. The bits are traveling at high speeds and removing pieces of usually scalding hot metal. The advantage of having a CNC milling machine is that it protects the machine operator.

## Turning

A lathe cutting material from a workpiece.

Turning is a metal cutting process for producing a cylindrical surface with a single point tool. The workpiece is rotated on a spindle and the cutting tool is fed into it radially, axially or both. Producing surfaces perpendicular to the workpiece axis is called facing. Producing surfaces using both radial and axial feeds is called profiling.

A *lathe* is a machine tool which spins a block or cylinder of material so that when abrasive, cutting, or deformation tools are applied to the workpiece, it can be shaped to produce an object which has rotational symmetry about an axis of rotation. Examples of objects that can be produced on a lathe include candlestick holders, crankshafts, camshafts, and bearing mounts.

Lathes have four main components: the bed, the headstock, the carriage, and the tailstock. The bed is a precise & very strong base which all of the other components rest upon for alignment. The headstock's spindle secures the workpiece with a chuck, whose jaws (usually three or four) are tightened around the piece. The spindle rotates at high speed, providing the energy to cut the material. While historically lathes were powered by belts from a line shaft, modern examples uses electric motors. The workpiece extends out of the spindle along the axis of rotation above the flat bed. The carriage is a platform that can be moved, precisely and independently parallel and perpendicular to the axis of rotation. A hardened cutting tool is held at the desired height (usually the middle of the workpiece) by the toolpost. The carriage is then moved around the rotating workpiece, and the cutting tool gradually removes material from the workpiece. The tailstock can be slid along the axis of rotation and then locked in place as necessary. It may hold centers to further secure the workpiece, or cutting tools driven into the end of the workpiece.

Other operations that can be performed with a single point tool on a lathe are:

Chamfering: Cutting an angle on the corner of a cylinder.

Parting: The tool is fed radially into the workpiece to cut off the end of a part.

Threading: A tool is fed along and across the outside or i side surface of rotating parts to produce external or internal threads.

Boring: A single-point tool is fed linearly and parallel to the axis of rotation to create a round hole.

Drilling: Feeding the drill into the workpiece axially.

Knurling: Uses a tool to produce a rough surface texture on the work piece. Frequently used to allow grip by hand on a metal part.

Modern computer numerical control (CNC) lathes and (CNC) machining centres can do secondary operations like milling by using driven tools. When driven tools are used the work piece stops rotating and the driven tool executes the machining operation with a rotating cutting tool. The CNC machines use x, y, and z coordinates in order to control the turning tools and produce the product. Most modern day CNC lathes are able to produce most turned objects in 3D.

Nearly all types of metal can be turned, although more time & specialist cutting tools are needed for harder workpieces.

## Threading

There are many threading processes including: cutting threads with a tap or die, thread milling, single-point thread cutting, thread rolling and forming, and thread grinding. A *tap* is used to cut a female thread on the inside surface of a pre-drilled hole, while a *die* cuts a male thread on a preformed cylindrical rod.

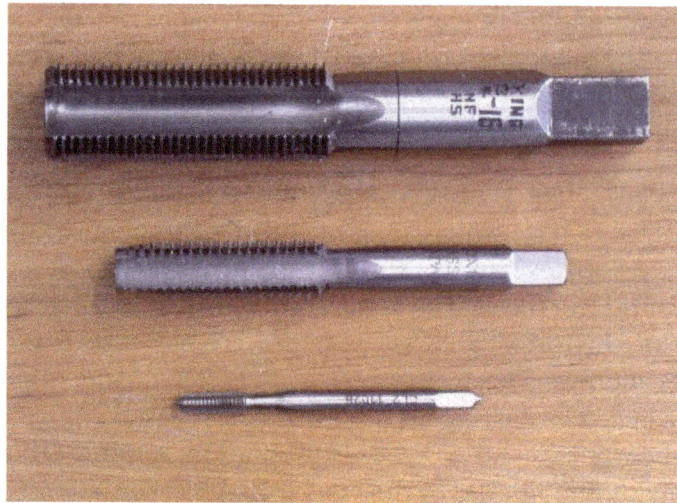

Three different types and sizes of taps.

## Grinding

A surface grinder

*Grinding* uses an abrasive process to remove material from the workpiece. A grinding machine is a machine tool used for producing very fine finishes, making very light cuts, or high precision forms using an abrasive wheel as the cutting device. This wheel can be made up of various sizes and types of stones, diamonds or inorganic materials.

The simplest grinder is a bench grinder or a hand-held angle grinder, for deburring parts or cutting metal with a zip-disc.

Grinders have increased in size and complexity with advances in time and technology. From the old days of a manual toolroom grinder sharpening endmills for a production shop, to today's 30000 RPM CNC auto-loading manufacturing cell producing jet turbines, grinding processes vary greatly.

Grinders need to be very rigid machines to produce the required finish. Some grinders are even used to produce glass scales for positioning CNC machine axis. The common rule is the machines used to produce scales be 10 times more accurate than the machines the parts are produced for.

In the past grinders were used for finishing operations only because of limitations of tooling. Modern grinding wheel materials and the use of industrial diamonds or other man-made coatings (cubic boron nitride) on wheel forms have allowed grinders to achieve excellent results in production environments instead of being relegated to the back of the shop.

Modern technology has advanced grinding operations to include CNC controls, high material removal rates with high precision, lending itself well to aerospace applications and high volume production runs of precision components.

## Filing

A file is an abrasive surface like this one that allows machinists to remove small, imprecise amounts of metal.

*Filing* is combination of grinding and saw tooth cutting using a file. Prior to the development of modern machining equipment it provided a relatively accurate means for the production of small parts, especially those with flat surfaces. The skilled use of a file allowed a machinist to work to fine tolerances and was the hallmark of the craft. Today filing is rarely used as a production technique in industry, though it remains as a common method of deburring.

## Other

Broaching is a machining operation used to cut keyways into shafts. Electron beam machining (EBM) is a machining process where high-velocity electrons are directed toward a work piece, creating heat and vaporizing the material. Ultrasonic machining uses ultrasonic vibrations to machine very hard or brittle materials.

## Joining Processes

Mig welding

## Welding

*Welding* is a fabrication process that joins materials, usually metals or thermoplastics, by causing coalescence. This is often done by melting the workpieces and adding a filler material to form a pool of molten material that cools to become a strong joint, but sometimes pressure is used in conjunction with heat, or by itself, to produce the weld.

Many different energy sources can be used for welding, including a gas flame, an electric arc, a laser, an electron beam, friction, and ultrasound. While often an industrial process, welding can be done in many different environments, including open air, underwater and in space. Regardless of location, however, welding remains dangerous, and precautions must be taken to avoid burns, electric shock, poisonous fumes, and overexposure to ultraviolet light.

## Brazing

*Brazing* is a joining process in which a filler metal is melted and drawn into a capillary formed by the assembly of two or more work pieces. The filler metal reacts metallurgically with the workpiece(s) and solidifies in the capillary, forming a strong joint. Unlike welding, the work piece is not melted. Brazing is similar to soldering, but occurs at temperatures in excess of 450 °C (842 °F). Brazing has the advantage of producing less thermal stresses than welding, and brazed assemblies tend to be more ductile than weldments because alloying elements can not segregate and precipitate.

Brazing techniques include, flame brazing, resistance brazing, furnace brazing, diffusion brazing, inductive brazing and vacuum brazing.

## Soldering

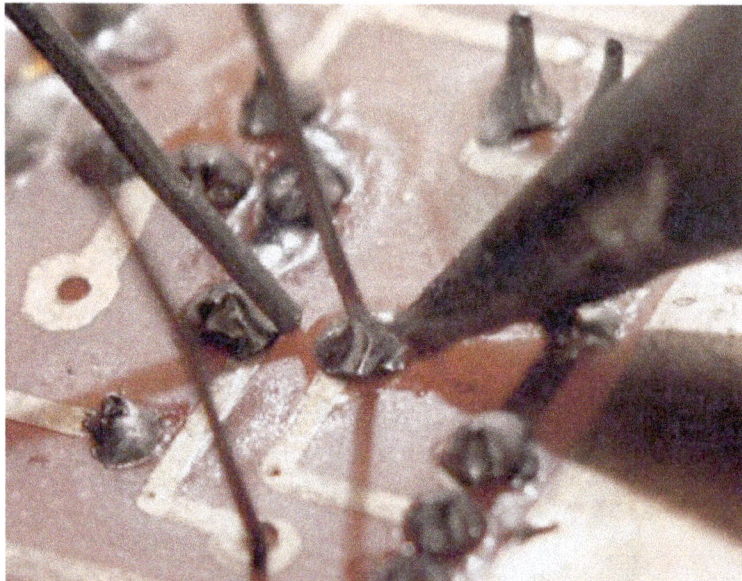

Soldering a printed circuit board.

*Soldering* is a joining process that occurs at temperatures below 450 °C (842 °F). It is similar to brazing in the way that a filler is melted and drawn into a capillary to form a join, although at a lower temperature. Because of this lower temperature and different alloys used as fillers, the metallurgical reaction between filler and work piece is minimal, resulting in a weaker joint.

## Riveting

*Riveting* is one of the most ancient metalwork joining processes. Its use has declined markedly during the second half of the 20th century, but it still retains important uses in industry and construction, and in artisan crafts such as jewellery, medieval armouring and Metal Couture into the 21st century. The earlier use of rivets is being superseded by improvements in welding and component fabrication techniques.

A rivet is essentially a two-headed and unthreaded bolt which holds two other pieces of metal together. Holes are drilled or punched through the two pieces of metal to be joined. The holes being aligned, a rivet is passed through the holes and permanent heads are formed onto the ends of the rivet utilizing hammers and forming dies (by either coldworking or hotworking). Rivets are commonly purchased with one head already formed.

When it is necessary to remove rivets, one of the rivet's heads is sheared off with a cold chisel. The rivet is then driven out with a hammer and punch.

## Associated Processes

While these processes are not primary metalworking processes, they are often performed before or after metalworking processes.

## Heat Treatment

Metals can be heat treated to alter the properties of strength, ductility, toughness, hardness or resistance to corrosion. Common heat treatment processes include annealing, precipitation hardening, quenching, and tempering. The annealing process softens the metal by allowing recovery of cold work and grain growth. Quenching can be used to harden alloy steels, or in precipitation hardenable alloys, to trap dissolved solute atoms in solution. Tempering will cause the dissolved alloying elements to precipitate, or in the case of quenched steels, improve impact strength and ductile properties.

Often, mechanical and thermal treatments are combined in what is known as thermo-mechanical treatments for better properties and more efficient processing of materials. These processes are common to high alloy special steels, super alloys and titanium alloys.

## Plating

Electroplating is a common surface-treatment technique. It involves bonding a thin layer of another metal such as gold, silver, chromium or zinc to the surface of the product. It is used to reduce corrosion as well as to improve the product's aesthetic appearance.

## Thermal Spraying

Thermal spraying techniques are another popular finishing option, and often have better high temperature properties than electroplated coatings.

# Casting (Metalworking)

Molten metal before casting

Casting iron in a sand mold

In metalworking, casting means a process, in which liquid metal is poured into a mold, that contains a hollow cavity of the desired shape, and then allowed to cool and solidify. The solidified part is also known as a casting, which is ejected or broken out of the mold to complete the process. Casting is most often used for making complex shapes that would be difficult or uneconomical to make by other methods.

Casting processes have been known for thousands of years, and widely used for sculpture, especially in bronze, jewellery in precious metals, and weapons and tools. Traditional techniques include lost-wax casting, plaster mold casting and sand casting.

The modern casting process is subdivided into two main categories: expendable and non-expendable casting. It is further broken down by the mold material, such as sand or metal, and pouring method, such as gravity, vacuum, or low pressure.

## Expendable Mold Casting

Expendable mold casting is a generic classification that includes sand, plastic, shell, plaster, and investment (lost-wax technique) moldings. This method of mold casting involves the use of temporary, non-reusable molds.

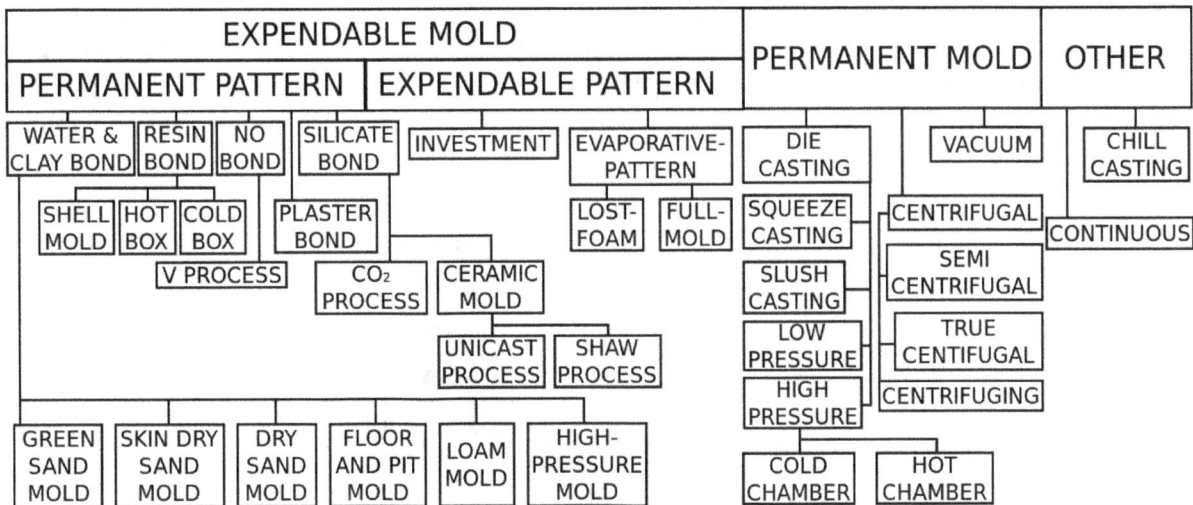

| EXPENDABLE MOLD | | | | | | PERMANENT MOLD | | OTHER |
|---|---|---|---|---|---|---|---|---|
| PERMANENT PATTERN | | | | EXPENDABLE PATTERN | | | | |
| WATER & CLAY BOND | RESIN BOND | NO BOND | SILICATE BOND | INVESTMENT | EVAPORATIVE-PATTERN | DIE CASTING | VACUUM | CHILL CASTING |
| SHELL MOLD | HOT BOX | COLD BOX | PLASTER BOND | LOST-FOAM / FULL-MOLD | | SQUEEZE CASTING | CENTRIFUGAL | CONTINUOUS |
| | V PROCESS | | $CO_2$ PROCESS | CERAMIC MOLD | | SLUSH CASTING | SEMI CENTRIFUGAL | |
| | | | | UNICAST PROCESS | SHAW PROCESS | LOW PRESSURE | TRUE CENTIFUGAL | |
| | | | | | | HIGH PRESSURE | CENTRIFUGING | |
| GREEN SAND MOLD | SKIN DRY SAND MOLD | DRY SAND MOLD | FLOOR AND PIT MOLD | LOAM MOLD | HIGH-PRESSURE MOLD | COLD CHAMBER | HOT CHAMBER | |

## Sand Casting

Sand casting is one of the most popular and simplest types of casting, and has been used for centuries. Sand casting allows for smaller batches than permanent mold casting and at a very reasonable cost. Not only does this method allow manufacturers to create products at a low cost, but there are other benefits to sand casting, such as very small-size operations. From castings that fit in the palm of your hand to train beds (one casting can create the entire bed for one rail car), it can all be done with sand casting. Sand casting also allows most metals to be cast depending on the type of sand used for the molds.

Sand casting requires a lead time of days, or even weeks sometimes, for production at high output rates (1–20 pieces/hr-mold) and is unsurpassed for large-part production. Green (moist) sand has almost no part weight limit, whereas dry sand has a practical part mass limit of 2,300–2,700 kg (5,100–6,000 lb). Minimum part weight ranges from 0.075–0.1 kg (0.17–0.22 lb). The sand is bonded together using clays, chemical binders, or polymerized oils (such as motor oil). Sand can be recycled many times in most operations and requires little maintenance.

## Plaster Mold Casting

Plaster casting is similar to sand casting except that plaster of paris is substituted for sand as a mold material. Generally, the form takes less than a week to prepare, after which a production rate of 1–10 units/hr·mold is achieved, with items as massive as 45 kg (99 lb) and as small as 30 g (1 oz) with very good surface finish and close tolerances. Plaster casting is an inexpensive alternative to other molding processes for complex parts due to the low cost of the plaster and its ability to produce near net shape castings. The biggest disadvantage is that it can only be used with low melting point non-ferrous materials, such as aluminium, copper, magnesium, and zinc.

## Shell Molding

Shell molding is similar to sand casting, but the molding cavity is formed by a hardened "shell" of sand instead of a flask filled with sand. The sand used is finer than sand casting sand and is mixed with a resin so that it can be heated by the pattern and hardened into a shell around the pattern. Because of the resin and finer sand, it gives a much finer surface finish. The process is easily automated and more precise than sand casting. Common metals that are cast include cast iron, aluminium, magnesium, and copper alloys. This process is ideal for complex items that are small to medium-sized.

## Investment Casting

An investment-cast valve cover

Investment casting (known as lost-wax casting in art) is a process that has been practiced for thousands of years, with the lost-wax process being one of the oldest known metal forming techniques. From 5000 years ago, when beeswax formed the pattern, to today's high technology waxes, refractory materials and specialist alloys, the castings ensure high-quality components are produced with the key benefits of accuracy, repeatability, versatility and integrity.

Investment casting derives its name from the fact that the pattern is invested, or surrounded, with a refractory material. The wax patterns require extreme care for they are not strong enough to withstand forces encountered during the mold making. One advantage of investment casting is that the wax can be reused.

The process is suitable for repeatable production of net shape components from a variety of different metals and high performance alloys. Although generally used for small castings, this process has been used to produce complete aircraft door frames, with steel castings of up to 300 kg and aluminium castings of up to 30 kg. Compared to other casting processes such as die casting or sand casting, it can be an expensive process. However, the components that can be produced using investment casting can incorporate intricate contours, and in most cases the components are cast near net shape, so require little or no rework once cast.

## Waste Molding of Plaster

A durable plaster intermediate is often used as a stage toward the production of a bronze sculpture or as a pointing guide for the creation of a carved stone. With the completion of a plaster, the work is more durable (if stored indoors) than a clay original which must be kept moist to avoid cracking. With the low cost plaster at hand, the expensive work of bronze casting or stone carving may be deferred until a patron is found, and as such work is considered to be a technical, rather than artistic process, it may even be deferred beyond the lifetime of the artist.

In waste molding a simple and thin plaster mold, reinforced by sisal or burlap, is cast over the original clay mixture. When cured, it is then removed from the damp clay, incidentally destroying the fine details in undercuts present in the clay, but which are now captured in the mold. The mold may then at any later time (but only once) be used to cast a plaster positive image, identical to the original clay. The surface of this plaster may be further refined and may be painted and waxed to resemble a finished bronze casting.

## Evaporative-Pattern Casting

This is a class of casting processes that use pattern materials that evaporate during the pour, which means there is no need to remove the pattern material from the mold before casting. The two main processes are lost-foam casting and full-mold casting.

## Lost-Foam Casting

Lost-foam casting is a type of evaporative-pattern casting process that is similar to investment casting except foam is used for the pattern instead of wax. This process takes advantage of the low boiling point of foam to simplify the investment casting process by removing the need to melt the wax out of the mold.

## Full-Mold Casting

Full-mold casting is an evaporative-pattern casting process which is a combination of sand casting and lost-foam casting. It uses an expanded polystyrene foam pattern which is then surrounded by sand, much like sand casting. The metal is then poured directly into the mold, which vaporizes the foam upon contact.

## Non-Expendable Mold Casting

Non-expendable mold casting differs from expendable processes in that the mold need not be reformed after each production cycle. This technique includes at least four different methods: permanent, die, centrifugal, and continuous casting. This form of casting also results in improved repeatability in parts produced and delivers Near Net Shape results.

The permanent molding process

## Permanent Mold Casting

Permanent mold casting is a metal casting process that employs reusable molds ("permanent molds"), usually made from metal. The most common process uses gravity to fill the mold. However, gas pressure or a vacuum are also used. A variation on the typical gravity casting process, called slush casting, produces hollow castings. Common casting metals are aluminum, magnesium, and copper alloys. Other materials include tin, zinc, and lead alloys and iron and steel are also cast in graphite molds. Permanent molds, while lasting more than one casting still have a limited life before wearing out.

## Die Casting

The die casting process forces molten metal under high pressure into mold cavities (which are machined into dies). Most die castings are made from nonferrous metals, specifically zinc, copper, and aluminium-based alloys, but ferrous metal die castings are possible. The die casting method is especially suited for applications where many small to medium-sized parts are needed with good detail, a fine surface quality and dimensional consistency.

## Semi-Solid Metal Casting

Semi-solid metal (SSM) casting is a modified die casting process that reduces or eliminates the residual porosity present in most die castings. Rather than using liquid metal as the feed material, SSM casting uses a higher viscosity feed material that is partially solid and partially liquid. A modified die casting machine is used to inject the semi-solid slurry into re-usable hardened steel dies. The high viscosity of the semi-solid metal, along with the use of controlled die filling conditions, ensures that the semi-solid metal fills the die in a non-turbulent manner so that harmful porosity can be essentially eliminated.

Used commercially mainly for aluminium and magnesium alloys, SSM castings can be heat treated to the T4, T5 or T6 tempers. The combination of heat treatment, fast cooling rates (from using un-coated steel dies) and minimal porosity provides excellent combinations of strength and ductility. Other advantages of SSM casting include the ability to produce complex shaped parts net shape, pressure tightness, tight dimensional tolerances and the ability to cast thin walls.

## Centrifugal Casting

In this process molten metal is poured in the mold and allowed to solidify while the mold is rotating. Metal is poured into the center of the mold at its axis of rotation. Due to centrifugal force the liquid metal is thrown out towards the periphery.

Centrifugal casting is both gravity- and pressure-independent since it creates its own force feed using a temporary sand mold held in a spinning chamber at up to 900 N. Lead time varies with the application. Semi- and true-centrifugal processing permit 30–50 pieces/hr-mold to be produced, with a practical limit for batch processing of approximately 9000 kg total mass with a typical per-item limit of 2.3–4.5 kg.

Industrially, the centrifugal casting of railway wheels was an early application of the method developed by the German industrial company Krupp and this capability enabled the rapid growth of the enterprise.

Small art pieces such as jewelry are often cast by this method using the lost wax process, as the forces enable the rather viscous liquid metals to flow through very small passages and into fine details such as leaves and petals. This effect is similar to the benefits from vacuum casting, also applied to jewelry casting.

## Continuous Casting

Continuous casting is a refinement of the casting process for the continuous, high-volume production of metal sections with a constant cross-section. Molten metal is poured into an open-ended, water-cooled mold, which allows a 'skin' of solid metal to form over the still-liquid centre, gradually solidifying the metal from the outside in. After solidification, the strand, as it is sometimes called, is continuously withdrawn from the mold. Predetermined lengths of the strand can be cut off by either mechanical shears or traveling oxyacetylene torches and transferred to further forming processes, or to a stockpile. Cast sizes can range from strip (a few millimeters thick by about five meters wide) to billets (90 to 160 mm square) to slabs (1.25 m wide by 230 mm thick). Sometimes, the strand may undergo an initial hot rolling process before being cut.

Continuous casting is used due to the lower costs associated with continuous production of a standard product, and also increased quality of the final product. Metals such as steel, copper, aluminum and lead are continuously cast, with steel being the metal with the greatest tonnages cast using this method.

## Terminology

Metal casting processes uses the following terminology:

- Pattern: An approximate duplicate of the final casting used to form the mold cavity.

- Molding material: The material that is packed around the pattern and then the pattern is removed to leave the cavity where the casting material will be poured.

- Flask: The rigid wood or metal frame that holds the molding material.

    o Cope: The top half of the pattern, flask, mold, or core.

    o Drag: The bottom half of the pattern, flask, mold, or core.

- Core: An insert in the mold that produces internal features in the casting, such as holes.

    o Core print: The region added to the pattern, core, or mold used to locate and support the core.

- Mold cavity: The combined open area of the molding material and core, where the metal is poured to produce the casting.

- Riser: An extra void in the mold that fills with molten material to compensate for shrinkage during solidification.

- Gating system: The network of connected channels that deliver the molten material to the mold cavities.

    o Pouring cup or pouring basin: The part of the gating system that receives the molten material from the pouring vessel.

    o Sprue: The pouring cup attaches to the sprue, which is the vertical part of the gating system. The other end of the sprue attaches to the runners.

    o Runners: The horizontal portion of the gating system that connects the sprues to the gates.

    o Gates: The controlled entrances from the runners into the mold cavities.

- Vents: Additional channels that provide an escape for gases generated during the pour.

- Parting line or parting surface: The interface between the cope and drag halves of the mold, flask, or pattern.

- Draft: The taper on the casting or pattern that allow it to be withdrawn from the mold

- Core box: The mold or die used to produce the cores.

Some specialized processes, such as die casting, use additional terminology.

## Theory

Casting is a solidification process, which means the solidification phenomenon controls most of the properties of the casting. Moreover, most of the casting defects occur during solidification, such as *gas porosity* and *solidification shrinkage*.

Solidification occurs in two steps: *nucleation* and *crystal growth*. In the nucleation stage solid particles form within the liquid. When these particles form their internal energy is lower than the surrounded liquid, which creates an energy interface between the two. The formation of the surface at this interface requires energy, so as nucleation occurs the material actually undercools, that is it cools below its freezing temperature, because of the extra energy required to form the interface surfaces. It then recalescences, or heats back up to its freezing temperature, for the crystal growth stage. Note that nucleation occurs on a pre-existing solid surface, because not as much energy is required for a partial interface surface, as is for a complete spherical interface surface. This can be advantageous because fine-grained castings possess better properties than coarse-grained castings. A fine grain structure can be induced by *grain refinement* or *inoculation*, which is the process of adding impurities to induce nucleation.

All of the nucleations represent a crystal, which grows as the heat of fusion is extracted from the liquid until there is no liquid left. The direction, rate, and type of growth can be controlled to maximize the properties of the casting. Directional solidification is when the material solidifies at one end and proceeds to solidify to the other end; this is the most ideal type of grain growth because it allows liquid material to compensate for shrinkage.

## Cooling Curves

Intermediate cooling rates from melt result in a dendritic microstructure. Primary and secondary dendrites can be seen in this image.

Cooling curves are important in controlling the quality of a casting. The most important part of the cooling curve is the *cooling rate* which affects the microstructure and properties. Generally speaking, an area of the casting which is cooled quickly will have a fine grain structure and an area which cools slowly will have a coarse grain structure. Below is an example cooling curve of a pure metal or eutectic alloy, with defining terminology.

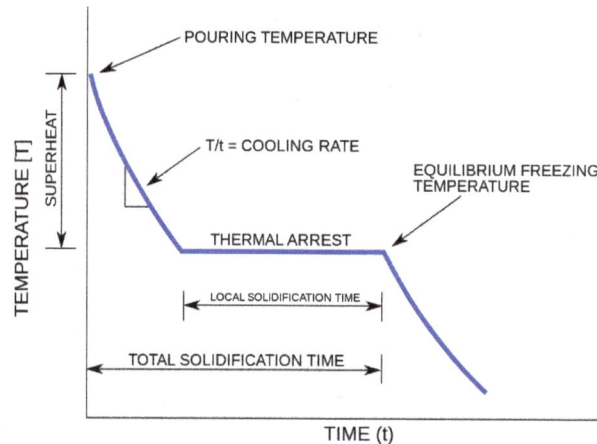

Note that before the thermal arrest the material is a liquid and after it the material is a solid; during the thermal arrest the material is converting from a liquid to a solid. Also, note that the greater the superheat the more time there is for the liquid material to flow into intricate details.

The above cooling curve depicts a basic situation with a pure alloy, however, most castings are of alloys, which have a cooling curve shaped as shown below.

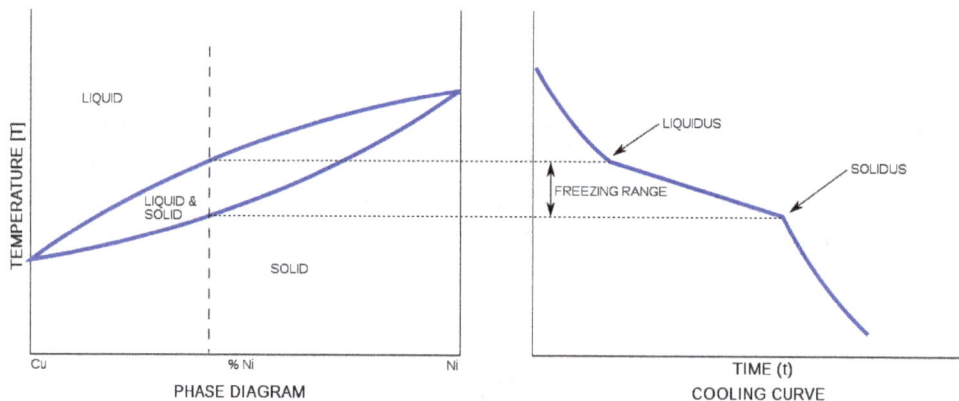

Note that there is no longer a thermal arrest, instead there is a freezing range. The freezing range corresponds directly to the liquidus and solidus found on the phase diagram for the specific alloy.

## Chvorinov's Rule

The local solidification time can be calculated using Chvorinov's rule, which is:

Where $t$ is the solidification time, $V$ is the volume of the casting, $A$ is the surface area of the casting that contacts the mold, $n$ is a constant, and $B$ is the mold constant. It is most useful in determining if a riser will solidify before the casting, because if the riser does solidify first then it is worthless.

## The Gating System

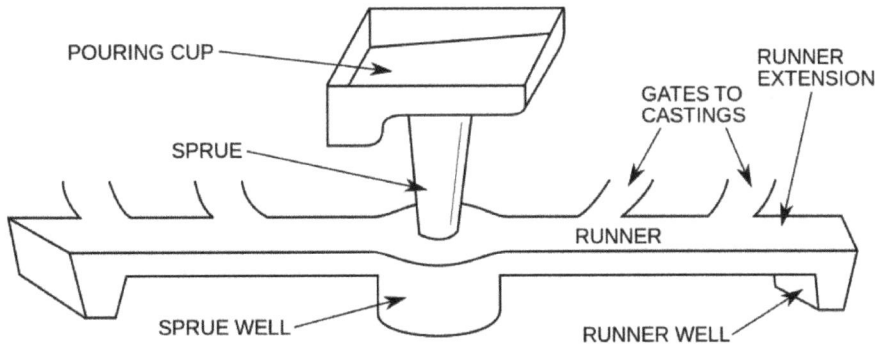

A simple gating system for a horizontal parting mold.

The gating system serves many purposes, the most important being conveying the liquid material to the mold, but also controlling shrinkage, the speed of the liquid, turbulence, and trapping dross. The gates are usually attached to the thickest part of the casting to assist in controlling shrinkage. In especially large castings multiple gates or runners may be required to introduce metal to more than one point in the mold cavity. The speed of the material is important because if the material is traveling too slowly it can cool before completely filling, leading to misruns and cold shuts. If the material is moving too fast then the liquid material can erode the mold and contaminate the final casting. The shape and length of the gating system can also control how quickly the material cools; short round or square channels minimize heat loss.

The gating system may be designed to minimize turbulence, depending on the material being cast. For example, steel, cast iron, and most copper alloys are turbulent insensitive, but aluminium and magnesium alloys are turbulent sensitive. The turbulent insensitive materials usually have a short and open gating system to fill the mold as quickly as possible. However, for turbulent sensitive materials short sprues are used to minimize the distance the material must fall when entering the mold. Rectangular pouring cups and tapered sprues are used to prevent the formation of a vortex as the material flows into the mold; these vortices tend to suck gas and oxides into the mold. A large sprue well is used to dissipate the kinetic energy of the liquid material as it falls down the sprue, decreasing turbulence. The *choke*, which is the smallest cross-sectional area in the gating system used to control flow, can be placed near the sprue well to slow down and smooth out the flow. Note that on some molds the choke is still placed on the gates to make separation of the part easier, but induces extreme turbulence. The gates are usually attached to the bottom of the casting to minimize turbulence and splashing.

The gating system may also be designed to trap dross. One method is to take advantage of the fact that some dross has a lower density than the base material so it floats to the top of the gating system. Therefore, long flat runners with gates that exit from the bottom of the runners can trap dross in the runners; note that long flat runners will cool the material more rapidly than round or square runners. For materials where the dross is a similar density to the base material, such as aluminium, *runner extensions* and *runner wells* can be advantageous. These take advantage of the fact that the dross is usually located at the beginning of the pour, therefore the runner is extended past the last gate(s) and the contaminates are contained in the wells. Screens or filters may also be used to trap contaminates.

It is important to keep the size of the gating system small, because it all must be cut from the casting and remelted to be reused. The efficiency, or *yield*, of a casting system can be calculated by dividing the weight of the casting by the weight of the metal poured. Therefore, the higher the number the more efficient the gating system/risers.

## Shrinkage

There are three types of shrinkage: *shrinkage of the liquid, solidification shrinkage* and *patternmaker's shrinkage*. The shrinkage of the liquid is rarely a problem because more material is flowing into the mold behind it. Solidification shrinkage occurs because metals are less dense as a liquid than a solid, so during solidification the metal density dramatically increases. Patternmaker's shrinkage refers to the shrinkage that occurs when the material is cooled from the solidification temperature to room temperature, which occurs due to thermal contraction.

## Solidification Shrinkage

| Solidification shrinkage of various metals | |
|---|---|
| **Metal** | **Percentage** |
| Aluminium | 6.6 |
| Copper | 4.9 |
| Magnesium | 4.0 or 4.2 |
| Zinc | 3.7 or 6.5 |
| Low carbon steel | 2.5–3.0 |
| High carbon steel | 4.0 |
| White cast iron | 4.0–5.5 |
| Gray cast iron | −2.5–1.6 |
| Ductile cast iron | −4.5–2.7 |

Most materials shrink as they solidify, but, as the table to the right shows, a few materials do not, such as gray cast iron. For the materials that do shrink upon solidification the type of shrinkage depends on how wide the freezing range is for the material. For materials with a narrow freezing range, less than 50 °C (122 °F), a cavity, known as a *pipe*, forms in the center of the casting, because the outer shell freezes first and progressively solidifies to the center. Pure and eutectic metals usually have narrow solidification ranges. These materials tend to form a *skin* in open air molds, therefore they are known as *skin forming alloys*. For materials with a wide freezing range, greater than 110 °C (230 °F), much more of the casting occupies the *mushy* or *slushy* zone (the temperature range between the solidus and the liquidus), which leads to small pockets of liquid trapped throughout and ultimately porosity. These castings tend to have poor ductility, toughness, and fatigue resistance. Moreover, for these types of materials to be fluid-tight a secondary operation is required to impregnate the casting with a lower melting point metal or resin.

For the materials that have narrow solidification ranges pipes can be overcome by designing the casting to promote directional solidification, which means the casting freezes first at the point farthest from the gate, then progressively solidifies towards the gate. This allows a continuous feed of liquid material to be present at the point of solidification to compensate for the shrinkage. Note that there is still a shrinkage void where the final material solidifies, but if designed properly this will be in the gating system or riser.

## Risers and Riser Aids

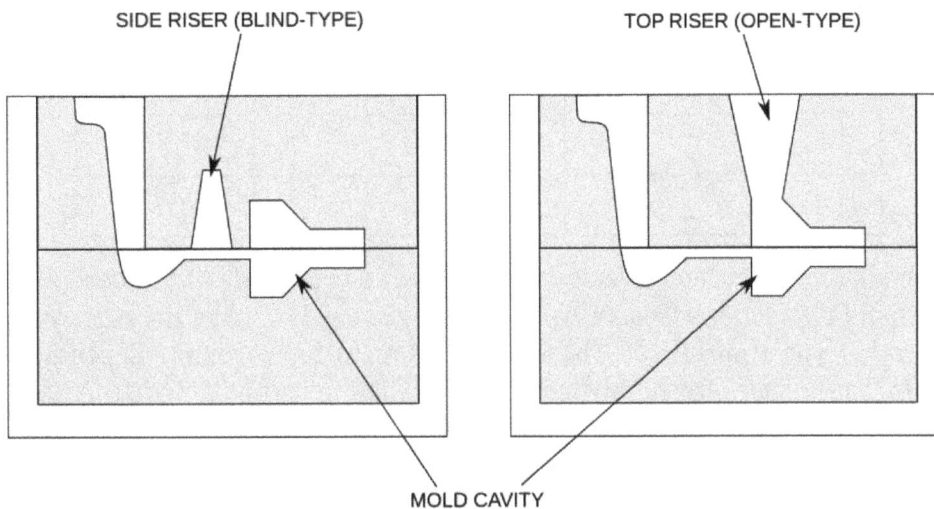

Different types of risers

Risers, also known as *feeders*, are the most common way of providing directional solidification. It supplies liquid metal to the solidifying casting to compensate for solidification shrinkage. For a riser to work properly the riser must solidify after the casting, otherwise it cannot supply liquid metal to shrinkage within the casting. Risers add cost to the casting because it lowers the *yield* of each casting; i.e. more metal is lost as scrap for each casting. Another way to promote directional solidification is by adding chills to the mold. A chill is any material which will conduct heat away from the casting more rapidly that the material used for molding.

Risers are classified by three criteria. The first is if the riser is open to the atmosphere, if it is then it is called an *open* riser, otherwise it is known as a *blind* type. The second criterion is where the riser is located; if it is located on the casting then it is known as a *top riser* and if it is located next to the casting it is known as a *side riser*. Finally, if riser is located on the gating system so that it fills after the molding cavity, it is known as a *live riser* or *hot riser*, but if the riser fills with materials that's already flowed through the molding cavity it is known as a *dead riser* or *cold riser*.

Riser aids are items used to assist risers in creating directional solidification or reducing the number of risers required. One of these items are *chills* which accelerate cooling in a certain part of the mold. There are two types: external and internal chills. External chills are masses of high-heat-capacity and high-thermal-conductivity material that are placed on an edge of the molding cavity. Internal chills are pieces of the same metal that is being poured, which are placed inside the mold cavity and become part of the casting. Insulating sleeves and toppings may also be installed around the riser cavity to slow the solidification of the riser. Heater coils may also be installed around or above the riser cavity to slow solidification.

## Patternmaker's Shrink

| Typical patternmaker's shrinkage of various metals | | |
|---|---|---|
| **Metal** | **Percentage** | **in/ft** |
| Aluminium | 1.0–1.3 | $\frac{1}{8}-\frac{5}{32}$ |
| Brass | 1.5 | $\frac{3}{16}$ |
| Magnesium | 1.0–1.3 | $\frac{1}{8}-\frac{5}{32}$ |
| Cast iron | 0.8–1.0 | $\frac{1}{10}-\frac{1}{8}$ |
| Steel | 1.5–2.0 | $\frac{3}{16}-\frac{1}{4}$ |

Shrinkage after solidification can be dealt with by using an oversized pattern designed specifically for the alloy used. *Contraction rules*, or *shrink rules*, are used to make the patterns oversized to compensate for this type of shrinkage. These rulers are up to 2.5% oversize, depending on the material being cast. These rulers are mainly referred to by their percentage change. A pattern made to match an existing part would be made as follows: First, the existing part would be measured using a standard ruler, then when constructing the pattern, the pattern maker would use a contraction rule, ensuring that the casting would contract to the correct size.

Note that patternmaker's shrinkage does not take phase change transformations into account. For example, eutectic reactions, martensitic reactions, and graphitization can cause expansions or contractions.

### Mold cavity

The mold cavity of a casting does not reflect the exact dimensions of the finished part due to a number of reasons. These modifications to the mold cavity are known as *allowances* and account for patternmaker's shrinkage, draft, machining, and distortion. In non-expendable processes, these allowances are imparted directly into the permanent mold, but in expendable mold processes they are imparted into the patterns, which later form the mold cavity. Note that for non-expendable molds an allowance is required for the dimensional change of the mold due to heating to operating temperatures.

For surfaces of the casting that are perpendicular to the parting line of the mold a draft must be included. This is so that the casting can be released in non-expendable processes or the pattern can be released from the mold without destroying the mold in expendable processes. The required draft angle depends on the size and shape of the feature, the depth of the mold cavity, how the part or pattern is being removed from the mold, the pattern or part material, the mold material, and the process type. Usually the draft is not less than 1%.

The machining allowance varies drastically from one process to another. Sand castings generally have a rough surface finish, therefore need a greater machining allowance, whereas die casting has a very fine surface finish, which may not need any machining tolerance. Also, the draft may provide enough of a machining allowance to begin with.

The distortion allowance is only necessary for certain geometries. For instance, U-shaped castings will tend to distort with the legs splaying outward, because the base of the shape can contract while the legs are constrained by the mold. This can be overcome by designing the mold cavity to slope the leg inward to begin with. Also, long horizontal sections tend to sag in the middle if ribs are not incorporated, so a distortion allowance may be required.

Cores may be used in expendable mold processes to produce internal features. The core can be of metal but it is usually done in sand.

## Filling

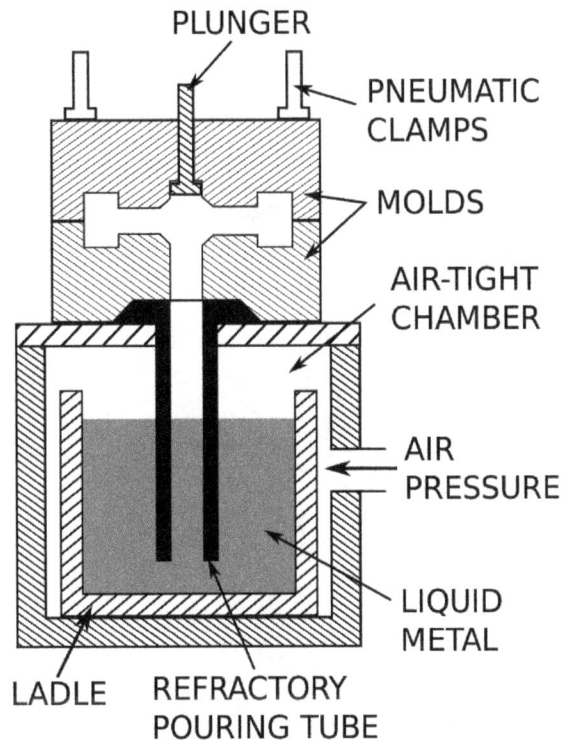

Schematic of the low-pressure permanent mold casting process

There are a few common methods for filling the mold cavity: *gravity*, *low-pressure*, *high-pressure*, and *vacuum*.

Vacuum filling, also known as *counter-gravity* filling, is more metal efficient than gravity pouring because less material solidifies in the gating system. Gravity pouring only has a 15 to 50% metal yield as compared to 60 to 95% for vacuum pouring. There is also less turbulence, so the gating system can be simplified since it does not have to control turbulence. Plus, because the metal is drawn from below the top of the pool the metal is free from dross and slag, as these are lower density (lighter) and float to the top of the pool. The pressure differential helps the metal flow into every intricacy of the mold. Finally, lower temperatures can be used, which improves the grain structure. The first patented vacuum casting machine and process dates to 1879.

Low-pressure filling uses 5 to 15 psig (35 to 100 kPag) of air pressure to force liquid metal up a feed tube into the mold cavity. This eliminates turbulence found in gravity casting and increases density, repeatability, tolerances, and grain uniformity. After the casting has solidified the pressure is released and any remaining liquid returns to the crucible, which increases yield.

### Tilt Filling

*Tilt filling*, also known as *tilt casting*, is an uncommon filling technique where the crucible is attached to the gating system and both are slowly rotated so that the metal enters the mold cavity with little turbulence. The goal is to reduce porosity and inclusions by limiting turbulence. For most uses tilt filling is not feasible because the following inherent problem: if the system is rotated slow enough to not induce turbulence, the front of the metal stream begins to solidify, which results in mis-runs. If the system is rotated faster then it induces turbulence, which defeats the purpose. Durville of France was the first to try tilt casting, in the 1800s. He tried to use it to reduce surface defects when casting coinage from aluminium bronze.

### Macrostructure

The grain macrostructure in ingots and most castings have three distinct regions or zones: the chill zone, columnar zone, and equiaxed zone. The image below depicts these zones.

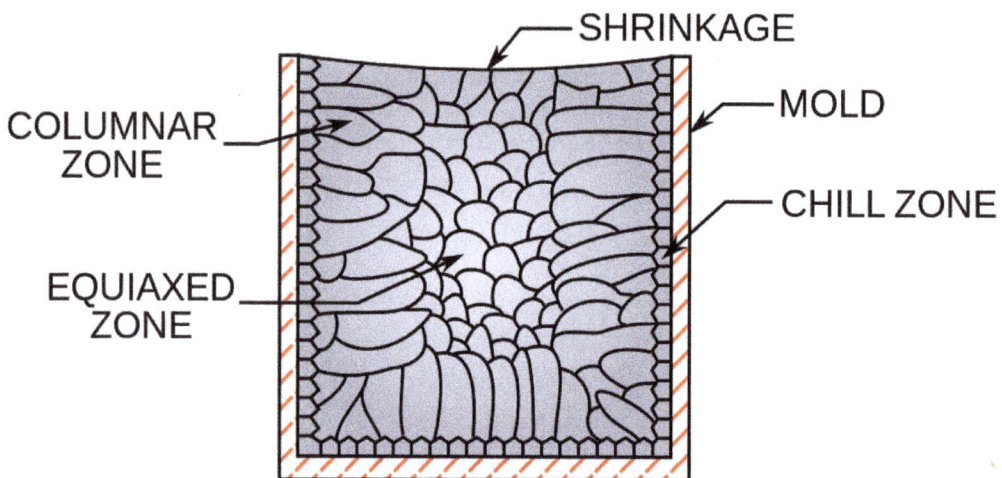

The chill zone is named so because it occurs at the walls of the mold where the wall *chills* the material. Here is where the nucleation phase of the solidification process takes place. As more heat is removed the grains grow towards the center of the casting. These are thin, long *columns* that are perpendicular to the casting surface, which are undesirable because they have anisotropic properties. Finally, in the center the equiaxed zone contains spherical, randomly oriented crystals. These are desirable because they have isotropic properties. The creation of this zone can be promoted by using a low pouring temperature, alloy inclusions, or inoculants.

### Inspection

Common inspection methods for steel castings are *magnetic particle testing* and *liquid penetrant testing*. Common inspection methods for aluminum castings are *radiography, ultrasonic testing,* and *liquid penetrant testing*.

## Defects

There are a number of problems that can be encountered during the casting process. The main types are: *gas porosity*, *shrinkage defects*, *mold material defects*, *pouring metal defects*, and *metallurgical defects*.

## Casting Process Simulation

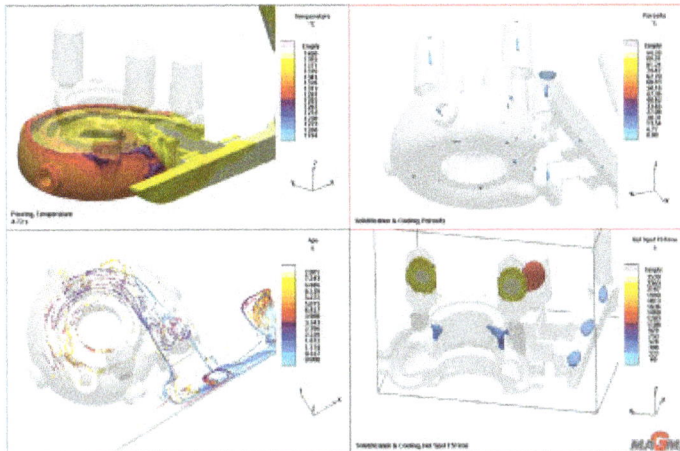

A high-performance software for the simulation of casting processes provides opportunities for an interactive or automated evaluation of results (here, for example, of mold filling and solidification, porosity and flow characteristics). Picture: Componenta B.V., The Netherlands)

Casting process simulation uses numerical methods to calculate cast component quality considering mold filling, solidification and cooling, and provides a quantitative prediction of casting mechanical properties, thermal stresses and distortion. Simulation accurately describes a cast component's quality up-front before production starts. The casting rigging can be designed with respect to the required component properties. This has benefits beyond a reduction in pre-production sampling, as the precise layout of the complete casting system also leads to energy, material, and tooling savings.

The software supports the user in component design, the determination of melting practice and casting methoding through to pattern and mold making, heat treatment, and finishing. This saves costs along the entire casting manufacturing route.

Casting process simulation was initially developed at universities starting from the early '70s, mainly in Europe and in the U.S., and is regarded as the most important innovation in casting technology over the last 50 years. Since the late '80s, commercial programs are available which make it possible for foundries to gain new insight into what is happening inside the mold or die during the casting process.

# Foundry

A foundry is a factory that produces metal castings. Metals are cast into shapes by melting them into a liquid, pouring the metal in a mold, and removing the mold material or casting after the metal has solidified as it cools. The most common metals processed are aluminium and cast iron. However, other metals, such as bronze, brass, steel, magnesium, and zinc, are also used to produce castings in foundries. In this process, parts of desired shapes and sizes can be formed.

*From Fra Burmeister og Wain's Iron Foundry*, by Peder Severin Krøyer, 1885.

## Process

A Foundryman, pictured by Daniel A. Wehrschmidt in 1899.

In metalworking, casting involves pouring liquid metal into a mold, which contains a hollow cavity of the desired shape, and then allowing it to cool and solidify. The solidified part is also known as a casting, which is ejected or broken out of the mold to complete the process. Casting is most often used for making complex shapes that would be difficult or uneconomical to make by other methods.

## Melting

Melting is performed in a furnace. Virgin material, external scrap, internal scrap, and alloying elements are used to charge the furnace. Virgin material refers to commercially pure forms of the primary metal used to form a particular alloy. Alloying elements are either pure forms of an alloying element, like electrolytic nickel, or alloys of limited composition, such as ferroalloys or master alloys. External scrap is material from other forming processes such as punching, forging, or machining. Internal scrap consists of gates, risers, defective castings, and other extraneous metal oddments produced within the facility.

Melting metal in a crucible for casting

A metal die casting robot in an industrial foundry

The process includes melting the charge, refining the melt, adjusting the melt chemistry and tapping into a transport vessel. Refining is done to remove deleterious gases and elements from the molten metal to avoid casting defects. Material is added during the melting process to bring the final chemistry within a specific range specified by industry and/or internal standards. Certain fluxes may be used to separate the metal from slag and/or dross and degassers are used to remove dissolved gas from metals that readily dissolve certain gasses. During the tap, final chemistry adjustments are made.

## Furnace

Several specialised furnaces are used to heat the metal. Furnaces are refractory-lined vessels that contain the material to be melted and provide the energy to melt it. Modern furnace types include electric arc furnaces (EAF), induction furnaces, cupolas, reverberatory, and crucible furnaces. Furnace choice is dependent on the alloy system quantities produced. For ferrous materials EAFs, cupolas, and induction furnaces are commonly used. Reverberatory and crucible furnaces are common for producing aluminium, bronze, and brass castings.

Furnace design is a complex process, and the design can be optimized based on multiple factors. Furnaces in foundries can be any size, ranging from small ones used to melt precious metals to furnaces weighing several tons, designed to melt hundreds of pounds of scrap at one time. They are designed according to the type of metals that are to be melted. Furnaces must also be designed based on the fuel being used to produce the desired temperature. For low temperature melting point alloys, such as zinc or tin, melting furnaces may reach around 500° C. Electricity, propane, or natural gas are usually used to achieve these temperatures. For high melting point alloys such as steel or nickel-based alloys, the furnace must be designed for temperatures over 1600° C. The fuel used to reach these high temperatures can be electricity (as employed in electric arc furnaces) or coke.

The majority of foundries specialize in a particular metal and have furnaces dedicated to these metals. For example, an iron foundry (for cast iron) may use a cupola, induction furnace, or EAF, while a steel foundry will use an EAF or induction furnace. Bronze or brass foundries use crucible furnaces or induction furnaces. Most aluminium foundries use either electric resistance or gas heated crucible furnaces or reverberatory furnaces.

## Degassing

Degassing is a process that may be required to reduce the amount of hydrogen present in a batch of molten metal. Gases can form in metal castings in one of two ways:

1.  by physical entrapment during the casting process or

2.  by chemical reaction in the cast material. Hydrogen is a common contaminant for most cast metals. It forms as a result of material reactions or from water vapor or machine lubricants. If the hydrogen concentration in the melt is too high, the resulting casting will be porous; the hydrogen will exit the molten solution, leaving minuscule air pockets, as the metal cools and solidifies. Porosity often seriously deteriorates the mechanical properties of the metal.

An efficient way of removing hydrogen from the melt is to bubble a dry, insoluble gas through the melt by purging or agitation. When the bubbles go up in the melt, they catch the dissolved hydrogen and bring it to the surface. Chlorine, nitrogen, helium and argon are often used to degas non-ferrous metals. Carbon monoxide is typically used for iron and steel.

There are various types of equipment that can measure the presence of hydrogen. Alternatively, the presence of hydrogen can be measured by determining the density of a metal sample.

In cases where porosity still remains present after the degassing process, porosity sealing can be accomplished through a process called metal impregnating.

# Mold Making

## Diagrams of Two Pattern Types

## DRAFT

A diagram of draft on a pattern

## UNDERCUT

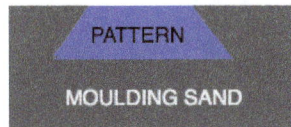

A diagram of an undercut in a mold

In the casting process a pattern is made in the shape of the desired part. Simple designs can be made in a single piece or solid pattern. More complex designs are made in two parts, called split patterns. A split pattern has a top or upper section, called a cope, and a bottom or lower section called a drag. Both solid and split patterns can have cores inserted to complete the final part shape. Cores are used to create hollow areas in the mold that would otherwise be impossible to achieve. Where the cope and drag separates is called the parting line.

When making a pattern it is best to taper the edges so that the pattern can be removed without breaking the mold. This is called draft. The opposite of draft is an undercut where there is part of the pattern under the mold material, making it impossible to remove the pattern without damaging the mold.

The pattern is made out of wax, wood, plastic, or metal. The molds are constructed by several different processes dependent upon the type of foundry, metal to be poured, quantity of parts to be produced, size of the casting, and complexity of the casting. These mold processes include:

- Sand casting — Green or resin bonded sand mold.

- Lost-foam casting — Polystyrene pattern with a mixture of ceramic and sand mold.

- Investment casting — Wax or similar sacrificial pattern with a ceramic mold.

- Ceramic mold casting — Plaster mold.

- V-process casting — Vacuum with thermoformed plastic to form sand molds. No moisture, clay or resin required.

- Die casting — Metal mold.

- Billet (ingot) casting — Simple mold for producing ingots of metal, normally for use in other foundries.

## Pouring

Bronze poured from a crucible into a mold, using the lost-wax casting process

In a foundry, molten metal is poured into molds. Pouring can be accomplished with gravity, or it may be assisted with a vacuum or pressurized gas. Many modern foundries use robots or automatic pouring machines to pour molten metal. Traditionally, molds were poured by hand using ladles.

## Shakeout

The solidified metal component is then removed from its mold. Where the mold is sand based, this can be done by shaking or tumbling. This frees the casting from the sand, which is still attached to the metal runners and gates — which are the channels through which the molten metal traveled to reach the component itself.

## Degating

Degating is the removal of the heads, runners, gates, and risers from the casting. Runners, gates, and risers may be removed using cutting torches, bandsaws, or ceramic cutoff blades. For some metal types, and with some gating system designs, the sprue, runners, and gates can be removed by breaking them away from the casting with a sledge hammer or specially designed knockout machinery. Risers must usually be removed using a cutting method but some newer methods of riser removal use knockoff machinery with special designs incorporated into the riser neck geometry that allow the riser to break off at the right place.

The gating system required to produce castings in a mold yields leftover metal — including heads, risers, and sprue (sometimes collectively called sprue) — that can exceed 50% of the metal required to pour a full mold. Since this metal must be remelted as salvage, the yield of a particular gating configuration becomes an important economic consideration when designing various gating schemes, to minimize the cost of excess sprue, and thus overall melting costs.

## Heat Treating

Heat treating is a group of industrial and metalworking processes used to alter the physical, and sometimes chemical, properties of a material. The most common application is metallurgical. Heat treatments are also used in the manufacture of many other materials, such as glass. Heat treatment involves the use of heating or chilling, normally to extreme temperatures, to achieve a desired result such as hardening or softening of a material. Heat treatment techniques include annealing, case hardening, precipitation strengthening, tempering, and quenching. It is noteworthy that while the term "heat treatment" applies only to processes where the heating and cooling are done for the specific purpose of altering properties intentionally, heating and cooling often occur incidentally during other manufacturing processes such as hot forming or welding.

## Surface Cleaning

After degating and heat treating, sand or other molding media may remain adhered to the casting. To remove any mold remnants, the surface is cleaned using a blasting process. This means a granular media will be propelled against the surface of the casting to mechanically knock away the adhering sand. The media may be blown with compressed air, or may be hurled using a shot wheel. The cleaning media strikes the casting surface at high velocity to dislodge the mold remnants (for example, sand, slag) from the casting surface. Numerous materials may be used to clean cast surfaces, including steel, iron, other metal alloys, aluminium oxides, glass beads, walnut shells, baking powder, and many others. The blasting media is selected to develop the color and reflectance of the cast surface. Terms used to describe this process include cleaning, bead blasting, and sand blasting. Shot peening may be used to further work-harden and finish the surface.

## Finishing

Modern foundry (circa 2000)

The final step in the process of casting usually involves grinding, sanding, or machining the component in order to achieve the desired dimensional accuracies, physical shape, and surface finish.

Removing the remaining gate material, called a gate stub, is usually done using a grinder or sander. These processes are used because their material removal rates are slow enough to control the amount of material being removed. These steps are done prior to any final machining.

After grinding, any surfaces that require tight dimensional control are machined. Many castings are machined in CNC milling centers. The reason for this is that these processes have better dimensional capability and repeatability than many casting processes. However, it is not uncommon today for castings to be used without machining.

A few foundries provide other services before shipping cast products to their customers. It is common to paint castings to prevent corrosion and improve visual appeal. Some foundries assemble castings into complete machines or sub-assemblies. Other foundries weld multiple castings or wrought metals together to form a finished product.

More and more, finishing processes are being performed by robotic machines, which eliminate the need for a human to physically grind or break parting lines, gating material, or feeders. Machines can reduce risk of injury to workers and lower costs for consumables — while also increasing productivity. They also limit the potential for human error and increase repeatability in the quality of grinding.

# Hot Working

A forge fire for hot working of metal

Hot working refers to processes where metals are plastically deformed above their recrystallization temperature. Being above the recrystallization temperature allows the material to recrystallize during deformation. This is important because recrystallization keeps the materials from strain hardening, which ultimately keeps the yield strength and hardness low and ductility high. This contrasts with cold working.

Many kinds of working, including rolling, forging, extrusion, and drawing, can be done with hot metal.

## Temperature

The lower limit of the hot working temperature is determined by its recrystallization temperature. As a guideline, the lower limit of the hot working temperature of a material is 60% its melting temperature (on an absolute temperature scale). The upper limit for hot working is determined by various factors, such as: excessive oxidation, grain growth, or an undesirable phase transformation. In practice materials are usually heated to the upper limit first to keep forming forces as low as possible and to maximize the amount of time available to hot work the workpiece.

The most important aspect of any hot working process is controlling the temperature of the workpiece. 90% of the energy imparted into the workpiece is converted into heat. Therefore, if the deformation process is quick enough the temperature of the workpiece should rise, however, this does not usually happen in practice. Most of the heat is lost through the surface of the workpiece into the cooler tooling. This causes temperature gradients in the workpiece, usually due to non-uniform cross-sections where the thinner sections are cooler than the thicker sections. Ultimately, this can lead to cracking in the cooler, less ductile surfaces. One way to minimize the problem is to heat the tooling. The hotter the tooling the less heat lost to it, but as the tooling temperature rises, the tool life decreases. Therefore the tooling temperature must be compromised; commonly, hot working tooling is heated to 500–850 °F (325–450 °C).

| Lower limit hot working temperature for various metals | |
|---|---|
| **Metal** | **Temperature** |
| Tin | Room temperature |
| Steel | 2,000 °F (1,090 °C) |
| Tungsten | 4,000 °F (2,200 °C) |

**Advantages & disadvantages**

The advantages are:

- Decrease in yield strength, therefore it is easier to work and uses less energy or force

- Increase in ductility

- Elevated temperatures increase diffusion which can remove or reduce chemical inhomogeneities

- Pores may reduce in size or close completely during deformation

- In steel, the weak, ductile, face-centered-cubic austenite microstructure is deformed instead of the strong body-centered-cubic ferrite microstructure found at lower temperatures

Usually the initial workpiece that is hot worked was originally cast. The microstructure of cast items does not optimize the engineering properties, from a microstructure standpoint. Hot working improves the engineering properties of the workpiece because it replaces the microstructure with one that has fine spherical shaped grains. These grains increase the strength, ductility, and toughness of the material.

The engineering properties can also be improved by reorienting the inclusions (impurities). In the cast state the inclusions are randomly oriented, which, when intersecting the surface, can be a propagation point for cracks. When the material is hot worked the inclusions tend to flow with the contour of the surface, creating *stringers*. As a whole the strings create a *flow structure*, where the properties are anisotropic (different based on direction). With the stringers oriented parallel to the surface it strengthens the workpiece, especially with respect to fracturing. The stringers act as "crack-arrestors" because the crack will want to propagate through the stringer and not along it.

The disadvantages are:

- Undesirable reactions between the metal and the surrounding atmosphere (scaling or rapid oxidation of the workpiece)

- Less precise tolerances due to thermal contraction and warping from uneven cooling

- Grain structure may vary throughout the metal for various reasons

- Requires a heating unit of some kind such as a gas or diesel furnace or an induction heater, which can be very expensive

# Rolling (Metalworking)

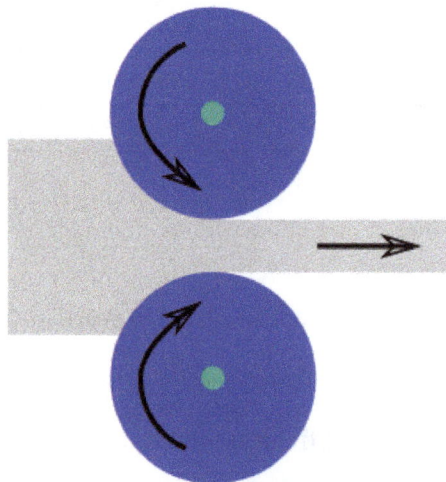

A rolling schematic

In metalworking, rolling is a metal forming process in which metal stock is passed through one or more pairs of rolls to reduce the thickness and to make the thickness uniform. The concept is similar to the rolling of dough. Rolling is classified according to the temperature of the metal rolled. If the temperature of the metal is above its recrystallization temperature, then the process is known as hot rolling. If the temperature of the metal is below its recrystallization temperature, the process is known as cold rolling. In terms of usage, hot rolling processes more tonnage than any other manufacturing process, and cold rolling processes the most tonnage out of all cold working processes. Roll stands holding pairs of rolls are grouped together into rolling mills that can quickly process metal, typically steel, into products such as structural steel (I-beams, angle stock, channel stock, and so on), bar stock, and rails. Most steel mills have rolling mill divisions that convert the semi-finished casting products into finished products.

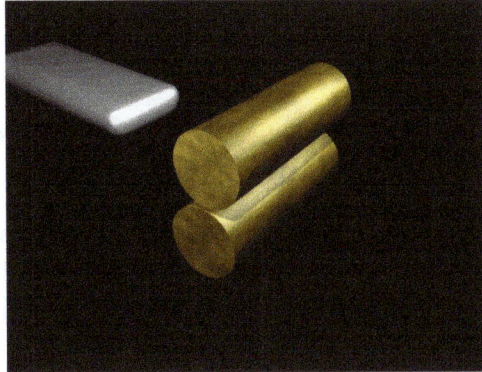

Rolling visualization. (Click on image to view animation.)

There are many types of rolling processes, including *ring rolling, roll bending, roll forming, profile rolling*, and *controlled rolling*.

## History

## Iron and Steel

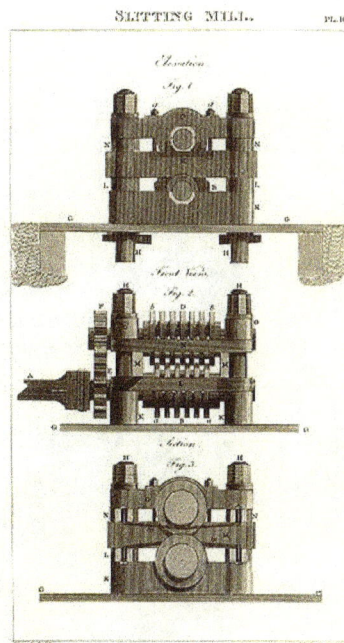

Slitting mill, 1813.

The discovery of the rolling mill in Europe maybe attributed to Leonardo da Vinci in his drawings. The earliest rolling mills in crude form but the same basic principles were found in Middle East and South Asia as early as 600 BCE. Earliest rolling mills were slitting mills, which were introduced from what is now Belgium to England in 1590. These passed flat bars between rolls to form a plate of iron, which was then passed between grooved rolls (slitters) to produce rods of iron. The first experiments at rolling iron for tinplate took place about 1670. In 1697, Major John Hanbury erected a mill at Pontypool to roll 'Pontypool plates'—blackplate. Later this began to be rerolled and tinned to make tinplate. The earlier production of plate iron in Europe had been in forges, not rolling mills.

The slitting mill was adapted to producing hoops (for barrels) and iron with a half-round or other sections by means that were the subject of two patents of c. 1679.,

Some of the earliest literature on rolling mills can be traced back to Christopher Polhem in 1761 in *Patriotista Testamente*, where he mentions rolling mills for both plate and bar iron. He also explains how rolling mills can save on time and labor because a rolling mill can produce 10 to 20 or more bars at the same time.

A patent was granted to Thomas Blockley of England in 1759 for the polishing and rolling of metals. Another patent was granted in 1766 to Richard Ford of England for the first tandem mill. A tandem mill is one in which the metal is rolled in successive stands; Ford's tandem mill was for hot rolling of wire rods.

## Other Metals

Rolling mills for lead seem to have existed by the late 17th century. Copper and brass were also rolled by the late 18th century.

## Modern Rolling

Properzi roller, Museo Nazionale della Scienza e della Tecnologia "Leonardo da Vinci", Milan

Modern rolling practice can be attributed to the pioneering efforts of Henry Cort of Funtley Iron Mills, near Fareham, England. In 1783 a patent was issued to Henry Cort for his use of grooved rolls for rolling iron bars. With this new design mills were able to produce 15 times more output per day than with a hammer. Although Cort was not the first to use grooved rolls, he was first to combine the use of many of the best features of various ironmaking and shaping processes known at the time. Thus modern writers have called him "father of modern rolling."

The first rail rolling mill was established by John Birkenshaw in 1820, where he produced fish bellied wrought iron rails in lengths of 15 to 18 feet. With the advancement of technology in rolling mills the size of rolling mills grew rapidly along with the size products being rolled. Example of this was at The Great Exhibition in 1851 a plate 20 feet long, 3 ½ feet wide, and 7/16 of inch thick, weighed 1,125 pounds was exhibited by the Consett Iron Company. Further evolution of the rolling mill came with the introduction of Three-high mills in 1853 used for rolling heavy sections.

## Hot and Cold Rolling

### Hot Rolling

A coil of hot-rolled steel

Hot rolling is a metalworking process that occurs above the recrystallization temperature of the material. After the grains deform during processing, they recrystallize, which maintains an equiaxed microstructure and prevents the metal from work hardening. The starting material is usually large pieces of metal, like semi-finished casting products, such as slabs, blooms, and billets. If these products came from a continuous casting operation the products are usually fed directly into the rolling mills at the proper temperature. In smaller operations the material starts at room temperature and must be heated. This is done in a gas- or oil-fired soaking pit for larger workpieces and for smaller workpieces induction heating is used. As the material is worked the temperature must be monitored to make sure it remains above the recrystallization temperature. To maintain a safety factor a *finishing temperature* is defined above the recrystallization temperature; this is usually 50 to 100 °C (90 to 180 °F) above the recrystallization temperature. If the temperature does drop below this temperature the material must be re-heated before more hot rolling.

Hot rolled metals generally have little directionality in their mechanical properties and deformation induced residual stresses. However, in certain instances non-metallic inclusions will impart some directionality and workpieces less than 20 mm (0.79 in) thick often have some directional properties. Also, non-uniform cooling will induce a lot of residual stresses, which usually occurs in shapes that have a non-uniform cross-section, such as I-beams. While the finished product is of good quality, the surface is covered in mill scale, which is an oxide that forms at high temperatures. It is usually removed via pickling or the smooth clean surface process, which reveals a smooth surface. Dimensional tolerances are usually 2 to 5% of the overall dimension.

Hot rolled mild steel seems to have a wider tolerance for amount of included carbon than does cold rolled steel, and is therefore more difficult for a blacksmith to use. Also for similar metals, hot rolled products seem to be less costly than cold-rolled ones.

Hot rolling is used mainly to produce sheet metal or simple cross sections, such as rail tracks. Other typical uses for hot rolled metal includes truck frames, automotive wheels, pipe and tubular, water heaters, agriculture equipment, strappings, stampings, compressor shells, railcar components, wheel rims, metal buildings, railroad hopper cars, doors, shelving, discs, guard rails, automotive clutch plates.

## Shape Rolling Design

Rolling mills are often divided into roughing, intermediate and finishing rolling cages. During shape rolling, an initial billet (round or square) with edge of diameter typically ranging between 100–140 mm is continuously deformed to produce a certain finished product with smaller cross section dimension and geometry. Different sequences can be adopted to produce a certain final product starting from a given billet. However, since each rolling mill is significantly expensive (up to 2 million of euros), a typical requirement is to contract the number or rolling passes. Different approaches have been achieved including, empirical knowledge, employment of numerical models and Artificial Intelligence techniques. Lambiase et al. validated a finite element model (FE) for predicting the final shape of a rolled bar in round-flat pass. one of the major concern when designing rolling mills is to reduce the number of passes; a possible solution to such requirement is represented by the slit pass also called split pass which divided an incoming bar in two or more sub-part thus virtually increasing the cross section reduction ratio per pass as reported by Lambiase. Another solution for reducing the number of passes in the rolling mills is the employment of automated systems for Roll Pass Design as that proposed by Lambiase and Langella. subsequently, Lambiase further developed an Automated System based on Artificial Intelligence and particularly an integrated system including an inferential engine based on Genetic Algorithms a knowledge database based on an Artificial Neural Network trained by a parametric Finite element model and to optimize and automatically design rolling mills.

## Cold Rolling

Cold rolling occurs with the metal below its recrystallization temperature (usually at room temperature), which increases the strength via strain hardening up to 20%. It also improves the surface finish and holds tighter tolerances. Commonly cold-rolled products include sheets, strips, bars, and rods; these products are usually smaller than the same products that are hot rolled. Because of the smaller size of the workpieces and their greater strength, as compared to hot rolled stock, four-high or cluster mills are used. Cold rolling cannot reduce the thickness of a workpiece as much as hot rolling in a single pass.

Cold-rolled sheets and strips come in various conditions: *full-hard, half-hard, quarter-hard*, and *skin-rolled*. Full-hard rolling reduces the thickness by 50%, while the others involve less of a reduction. Skin-rolling, also known as a *skin-pass*, involves the least amount of reduction: 0.5-1%. It is used to produce a smooth surface, a uniform thickness, and reduce the yield point phenomenon (by preventing Lüders bands from forming in later processing). It locks dislocations at the surface and thereby reduces the possibility of formation of Lüders bands. To avoid the formation of Lüders bands it is necessary to create substantial density of unpinned dislocations in ferrite matrix. It is also used to break up the spangles in galvanized steel. Skin-rolled stock is usually used in subsequent cold-working processes where good ductility is required.

Other shapes can be cold-rolled if the cross-section is relatively uniform and the transverse dimension is relatively small. Cold rolling shapes requires a series of shaping operations, usually along the lines of sizing, breakdown, roughing, semi-roughing, semi-finishing, and finishing.

If processed by a blacksmith, the smoother, more consistent, and lower levels of carbon encapsulated in the steel makes it easier to process, but at the cost of being more expensive.

Typical uses for cold-rolled steel include metal furniture, desks, filing cabinets, tables, chairs, motorcycle exhaust pipes, computer cabinets and hardware, home appliances and components, shelving, lighting fixtures, hinges, tubing, steel drums, lawn mowers, electronic cabinetry, water heaters, metal containers, and a variety of construction-related products.

## Processes

## Roll Bending

Roll bending

Roll bending produces a cylindrical shaped product from plate or steel metals .

## Roll Forming

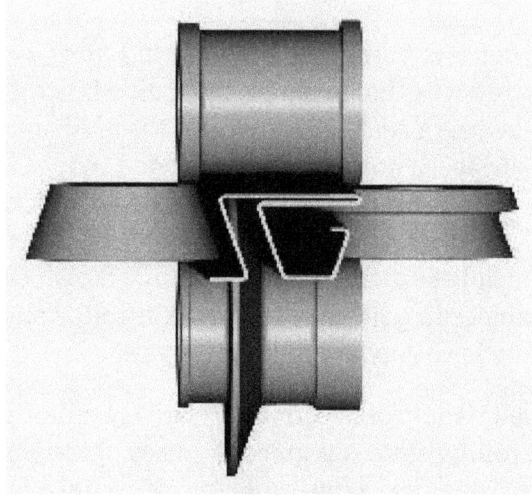

Roll forming

Roll forming, roll bending or plate rolling is a continuous bending operation in which a long strip of metal (typically coiled steel) is passed through consecutive sets of rolls, or stands, each performing only an incremental part of the bend, until the desired cross-section profile is obtained. Roll forming is ideal for producing parts with long lengths or in large quantities. There are 3 main processes: 4 rollers, 3 rollers and 2 rollers, each of which has as different advantages according to the desired specifications of the output plate.

## Flat Rolling

Flat rolling is the most basic form of rolling with the starting and ending material having a rectangular cross-section. The material is fed in between two *rollers*, called *working rolls*, that rotate in opposite directions. The gap between the two rolls is less than the thickness of the starting material, which causes it to deform. The decrease in material thickness causes the material to elongate. The friction at the interface between the material and the rolls causes the material to be pushed through. The amount of deformation possible in a single pass is limited by the friction between the rolls; if the change in thickness is too great the rolls just slip over the material and do not draw it in. The final product is either sheet or plate, with the former being less than 6 mm (0.24 in) thick and the latter greater than; however, heavy plates tend to be formed using a press, which is termed *forming*, rather than rolling.

Often the rolls are heated to assist in the workability of the metal. Lubrication is often used to keep the workpiece from sticking to the rolls. To fine-tune the process, the speed of the rolls and the temperature of the rollers are adjusted.

h is sheet metal with a thickness less than 200 μm (0.0079 in). The rolling is done in a *cluster mill* because the small thickness requires a small diameter rolls. To reduce the need for small rolls *pack rolling* is used, which rolls multiple sheets together to increase the effective starting thickness. As the foil sheets come through the rollers, they are trimmed and slitted with circular or razor-like knives. Trimming refers to the edges of the foil, while slitting involves cutting it into several sheets. Aluminum foil is the most commonly produced product via pack rolling. This is evident from the two different surface finishes; the shiny side is on the roll side and the dull side is against the other sheet of foil.

## Ring Rolling

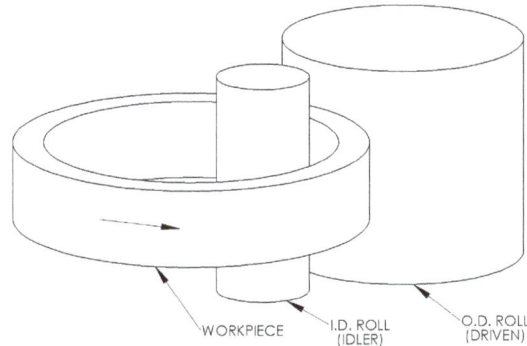

A schematic of ring rolling

Ring rolling is a specialized type of hot rolling that **increases** the diameter of a ring. The starting material is a thick-walled ring. This workpiece is placed between two rolls, an inner *idler roll* and a *driven roll*, which presses the ring from the outside. As the rolling occurs the wall thickness decreases as the diameter increases. The rolls may be shaped to form various cross-sectional shapes. The resulting grain structure is circumferential, which gives better mechanical properties. Diameters can be as large as 8 m (26 ft) and face heights as tall as 2 m (79 in). Common applications include bearings, gears, rockets, turbines, airplanes, pipes, and pressure vessels.

## Structural Shape rolling

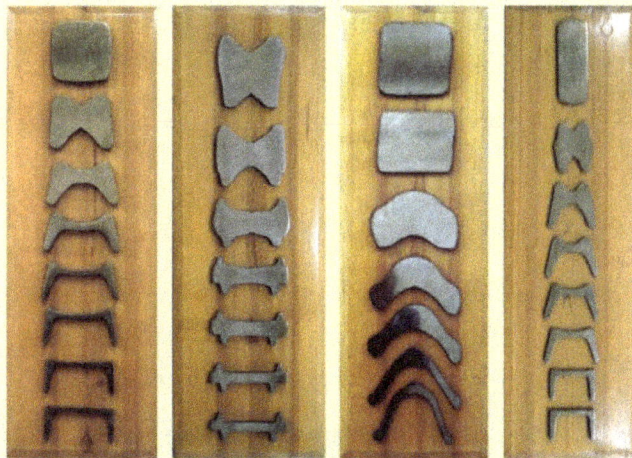

Cross-sections of continuously rolled structural shapes, showing the change induced by each rolling mill.

## Controlled Rolling

*Controlled rolling* is a type of thermomechanical processing which integrates controlled deformation and heat treating. The heat which brings the workpiece above the recrystallization temperature is also used to perform the heat treatments so that any subsequent heat treating is unnecessary. Types of heat treatments include the production of a fine grain structure; controlling the nature, size, and distribution of various transformation products (such as ferrite, austenite, pearlite, bainite, and martensite in steel); inducing precipitation hardening; and, controlling the toughness. In order to achieve this the entire process must be closely monitored and controlled. Common variables in controlled rolling include the starting material composition and structure, deformation levels, temperatures at various stages, and cool-down conditions. The benefits of controlled rolling include better mechanical properties and energy savings.

## Forge Rolling

Forge rolling is a longitudinal rolling process to reduce the cross-sectional area of heated bars or billets by leading them between two contrary rotating roll segments. The process is mainly used to provide optimized material distribution for subsequent die forging processes. Owing to this a better material utilization, lower process forces and better surface quality of parts can be achieved in die forging processes.

Basically any forgeable metal can also be forge-rolled. Forge rolling is mainly used to preform long-scaled billets through targeted mass distribution for parts such as crankshafts, connection rods, steering knuckles and vehicle axles. Narrowest manufacturing tolerances can only partially be achieved by forge rolling. This is the main reason why forge rolling is rarely used for finishing, but mainly for preforming.

Characteristics of forge rolling:

- high productivity and high material utilization
- good surface quality of forge-rolled workpieces
- extended tool life-time
- small tools and low tool costs
- improved mechanical properties due to optimized grain flow compared to exclusively die forged workpieces

## Mills

A *rolling mill*, also known as a *reduction mill* or *mill*, has a common construction independent of the specific type of rolling being performed:

Rolling mills

Rolling mill for cold rolling metal sheet like this piece of brass sheet

- Work rolls

- Backup rolls - are intended to provide rigid support required by the working rolls to prevent bending under the rolling load

- Rolling balance system - to ensure that the upper work and back up rolls are maintained in proper position relative to lower rolls

- Roll changing devices - use of an overhead crane and a unit designed to attach to the neck of the roll to be removed from or inserted into the mill.

- Mill protection devices - to ensure that forces applied to the backup roll chocks are not of such a magnitude to fracture the roll necks or damage the mill housing

- Roll cooling and lubrication systems

- Pinions - gears to divide power between the two spindles, rotating them at the same speed but in different directions

- Gearing - to establish desired rolling speed

- Drive motors - rolling narrow foil product to thousands of horsepower

- Electrical controls - constant and variable voltages applied to the motors

- Coilers and uncoilers - to unroll and roll up coils of metal

Slabs are the feed material for hot strip mills or plate mills and blooms are rolled to billets in a billet mill or large sections in a structural mill. The output from a strip mill is coiled and, subsequently, used as the feed for a cold rolling mill or used directly by fabricators. Billets, for re-rolling, are subsequently rolled in either a merchant, bar or rod mill. Merchant or bar mills produce a variety of shaped products such as angles, channels, beams, rounds (long or coiled) and hexagons.

## Configurations

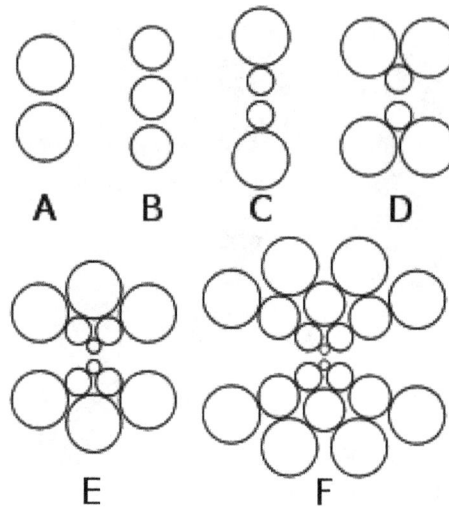

Various rolling configurations. Key: A. 2-high B. 3-high C. 4-high D. 6-high E. 12-high cluster & F. 20-high Sendzimir Mill cluster

Mills are designed in different types of configurations, with the most basic being a *two-high non-reversing*, which means there are two rolls that only turn in one direction. The *two-high reversing* mill has rolls that can rotate in both directions, but the disadvantage is that the rolls must be stopped, reversed, and then brought back up to rolling speed between each pass. To resolve this, the *three-high* mill was invented, which uses three rolls that rotate in one direction; the metal is fed through two of the rolls and then returned through the other pair. The disadvantage to this system is the workpiece must be lifted and lowered using an elevator. All of these mills are usually used for primary rolling and the roll diameters range from 60 to 140 cm (24 to 55 in).

To minimize the roll diameter a *four-high* or *cluster* mill is used. A small roll diameter is advantageous because less roll is in contact with the material, which results in a lower force and power requirement. The problem with a small roll is a reduction of stiffness, which is overcome using *backup rolls*. These backup rolls are larger and contact the back side of the smaller rolls. A four-high mill has four rolls, two small and two large. A cluster mill has more than 4 rolls, usually in three tiers. These types of mills are commonly used to hot roll wide plates, most cold rolling applications, and to roll foils.

Historically mills were classified by the product produced:

- Blooming, cogging and slabbing mills, being the preparatory mills to rolling finished rails, shapes or plates, respectively. If reversing, they are from 34 to 48 inches in diameter, and if three-high, from 28 to 42 inches in diameter.

- Billet mills, three-high, rolls from 24 to 32 inches in diameter, used for the further reduction of blooms down to 1.5x1.5-inch billets, being the preparatory mills for the bar and rod

- Beam mills, three-high, rolls from 28 to 36 inches in diameter, for the production of heavy beams and channels 12 inches and over.

- Rail mills with rolls from 26 to 40 inches in diameter.

- Shape mills with rolls from 20 to 26 inches in diameter, for smaller sizes of beams and channels and other structural shapes.

- Merchant bar mills with rolls from 16 to 20 inches in diameter.

- Small merchant bar mills with finishing rolls from 8 to 16 inches in diameter, generally arranged with a larger size roughing stand.

- Rod and wire mills with finishing rolls from 8 to 12 inches in diameter, always arranged with larger size roughing stands.

- Hoop and cotton tie mills, similar to small merchant bar mills.

- Armour plate mills with rolls from 44 to 50 inches in diameter and 140 to 180-inch body.

- Plate mills with rolls from 28 to 44 inches in diameter.

- Sheet mills with rolls from 20 to 32 inches in diameter.

- Universal mills for the production of square-edged or so-called universal plates and various wide flanged shapes by a system of vertical and horizontal rolls.

## Tandem Mill

A tandem mill is a special type of modern rolling mill where rolling is done in one pass. In a traditional rolling mill rolling is done in several passes, but in tandem mill there are several *stands* (>=2 stands) and reductions take place successively. The number of stands ranges from 2 to 18. Tandem mills can be either of hot or cold rolling mill types.

## Defects

In hot rolling, if the temperature of the workpiece is not uniform the flow of the material will occur more in the warmer parts and less in the cooler. If the temperature difference is great enough cracking and tearing can occur.

## Flatness and Shape

In a flat metal workpiece, the flatness is a descriptive attribute characterizing the extent of the geometric deviation from a reference plane. The deviation from complete flatness is the direct result of the workpiece relaxation after hot or cold rolling, due to the internal stress pattern caused by the non-uniform transversal compressive action of the rolls and the uneven geometrical properties of the entry material. The transverse distribution of differential strain/elongation-induced stress with respect to the material's average applied stress is commonly referenced to as shape. Due to the strict relationship between shape and flatness, these terms can be used in an interchangeable manner. In the case of metal strips and sheets, the flatness reflects the differential fiber elongation across the width of the workpiece. This property must be subject to an accurate feedback-based control in order to guarantee the machinability of the metal sheets in the final transformation processes. Some technological details about the feedback control of flatness are given in.

## Profile

Profile is made up of the measurements of crown and wedge. Crown is the thickness in the center as compared to the average thickness at the edges of the workpiece. Wedge is a measure of the thickness at one edge as opposed to the other edge. Both may be expressed as absolute measurements or as relative measurements. For instance, one could have 2 mil of crown (the center of the workpiece is 2 mil thicker than the edges), or one could have 2% crown (the center of the workpiece is 2% thicker than the edges).

It is typically desirable to have some crown in the workpiece as this will cause the workpiece to tend to pull to the center of the mill, and thus will run with higher stability.

## Flatness

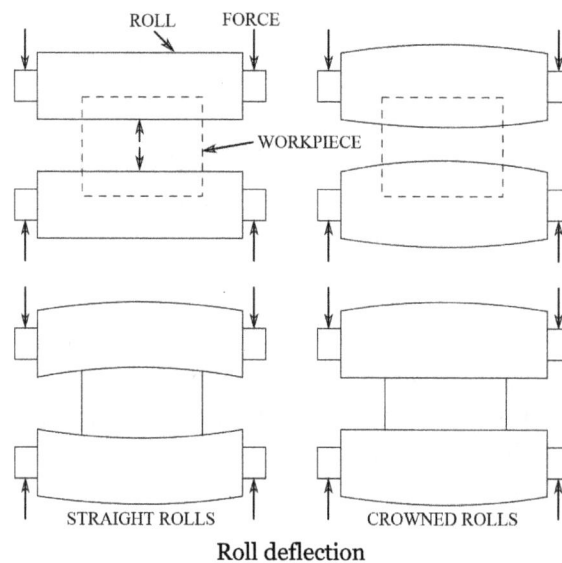

Roll deflection

Maintaining a uniform gap between the rolls is difficult because the rolls deflect under the load required to deform the workpiece. The deflection causes the workpiece to be thinner on the edges and thicker in the middle. This can be overcome by using a crowned roller (parabolic crown), however the crowned roller will only compensate for one set of conditions, specifically the material, temperature, and amount of deformation.

Other methods of compensating for roll deformation include continual varying crown (CVC), pair cross rolling, and work roll bending. CVC was developed by SMS-Siemag AG and involves grinding a third order polynomial curve into the work rolls and then shifting the work rolls laterally, equally, and opposite to each other. The effect is that the rolls will have a gap between them that is parabolic in shape, and will vary with lateral shift, thus allowing for control of the crown of the rolls dynamically. Pair cross rolling involves using either flat or parabolically crowned rolls, but shifting the ends at an angle so that the gap between the edges of the rolls will increase or decrease, thus allowing for dynamic crown control. Work roll bending involves using hydraulic cylinders at the ends of the rolls to counteract roll deflection.

Another way to overcome deflection issues is by decreasing the load on the rolls, which can be done by applying a longitudinal force; this is essentially drawing. Other method of decreasing roll deflection include increasing the elastic modulus of the roll material and adding back-up supports to the rolls.

The different classifications for flatness defects are:

- Symmetrical edge wave - the edges on both sides of the workpiece are "wavy" due to the material at the edges being longer than the material in the center.

- Asymmetrical edge wave - one edge is "wavy" due to the material at one side being longer than the other side.

- Center buckle - The center of the strip is "wavy" due to the strip in the center being longer than the strip at the edges.

- Quarter buckle - This is a rare defect where the fibers are elongated in the quarter regions (the portion of the strip between the center and the edge). This is normally attributed to using excessive roll bending force since the bending force may not compensate for the roll deflection across the entire length of the roll.

It is important to note that one could have a flatness defect even with the workpiece having the same thickness across the width. Also, one could have fairly high crown or wedge, but still produce material that is flat. In order to produce flat material, the material must be reduced by the same percentage across the width. This is important because mass flow of the material must be preserved, and the more a material is reduced, the more it is elongated. If a material is elongated in the same manner across the width, then the flatness coming into the mill will be preserved at the exit of the mill.

## Draught

The difference between the thickness of initial and rolled metal piece is called Draught. Thus if $t_0$ is initial thickness and $t_f$ is final thickness, then the draught $d$ is given by

$$d = t_0 - t_f$$

The maximum draught that can be achieved via rollers of radius $R$ with coefficient of static friction $f$ between the roller and the metal surface is given by

$$d_{max} = f^2 R$$

This is the case when the frictional force on the metal from inlet contact matches the negative force from the exit contact.

## Surface Defect Types

There are six types of surface defects:

Lap

> This type of defect occurs when a corner or fin is folded over and rolled but not welded into the metal. They appear as seams across the surface of the metal.

Mill-shearing

>These defects occur as a feather-like lap.

Rolled-in scale

>This occurs when mill scale is rolled into metal.

Scabs

>These are long patches of loose metal that have been rolled into the surface of the metal.

Seams

>They are open, broken lines that run along the length of the metal and caused by the presence of scale as well as due to pass roughness of Roughing mill.

Slivers

>Prominent surface ruptures.

## Surface Defect Remediation

Many surface defects can be scarfed off the surface of semi-finished rolled products before further rolling. Methods of scarfing have included hand-chipping with chisels (18th and 19th centuries); powered chipping and grinding with air chisels and grinders; burning with an oxy-fuel torch, whose gas pressure blows away the metal or slag melted by the flame; and laser scarfing.

# Cold-Formed Steel

Cold-formed steel building

*Cold-formed steel* (*CFS*) is the common term for products made by rolling or pressing steel into semi-finished or finished goods at relatively low temperatures (cold working). Cold-formed steel goods are created by the working of steel billet, bar, or sheet using stamping, rolling (including roll forming), or presses to deform it into a usable product. Cold-worked steel products, such as cold-rolled steel (CRS) bar stock and sheet, are commonly used in all areas of manufacturing of durable goods, such as appliances or automobiles, but the phrase *cold-formed steel* is most prevalently used to describe construction materials. The use of cold-formed steel construction materials has become more and more popular since its initial introduction of codified standards in 1946. In the construction industry both structural and non-structural elements are created from thin gauges of sheet steel. These building materials encompass columns, beams, joists, studs, floor decking, built-up sections and other components. Cold-formed steel construction materials differ from other steel construction materials known as hot-rolled steel. The manufacturing of cold-formed steel products occurs at room temperature using rolling or pressing. The strength of elements used for design is usually governed by buckling. The construction practices are more similar to timber framing using screws to assemble stud frames.

Cold-formed steel members have been used in buildings, bridges, storage racks, grain bins, car bodies, railway coaches, highway products, transmission towers, transmission poles, drainage facilities, various types of equipment and others. These types of sections are cold-formed from steel sheet, strip, plate, or flat bar in roll forming machines, by press brake (machine press) or bending operations. The material thicknesses for such thin-walled steel members usually range from 0.0147 in. (0.373 mm) to about ¼ in. (6.35 mm). Steel plates and bars as thick as 1 in. (25.4 mm) can also be cold-formed successfully into structural shapes (AISI, 2007b).

## History of Cold-Formed Steel

The use of cold-formed steel members in building construction began in the 1850s in both the United States and Great Britain. In the 1920s and 1930s, acceptance of cold-formed steel as a construction material was still limited because there was no adequate design standard and limited information on material use in building codes. One of the first documented uses of cold-formed steel as a building material is the Virginia Baptist Hospital , constructed around 1925 in Lynchburg, Virginia. The walls were load bearing masonry, but the floor system was framed with double back-to-back cold-formed steel lipped channels. According to Chuck Greene, P.E of Nolen Frisa Associates , the joists were adequate to carry the initial loads and spans, based on current analysis techniques. Greene engineered a recent renovation to the structure and said that for the most part, the joists are still performing well. A site observation during this renovation confirmed that "these joists from the 'roaring twenties' are still supporting loads, over 80 years later!" In the 1940s, Lustron Homes built and sold almost 2500 steel-framed homes, with the framing, finishes, cabinets and furniture made from cold-formed steel.

## History of AISI Design Standards

Design standards for hot-rolled steel were adopted in 1930s, but were not applicable to cold–formed sections because of their relatively thin steel walls which were susceptible to buckling. Cold-formed steel members maintain a constant thickness around their cross-section, whereas hot-rolled shapes typically exhibit tapering or fillets. Cold-formed steel allowed for shapes which differed greatly from the classical hot-rolled shapes. The material was easily workable; it could be deformed into many possible shapes. Even a small change in the geometry  created significant changes in the strength characteristics of the section. It was necessary to establish some minimum requirements and laws to control the buckling and strength characteristics. Also it was observed that the thin walls underwent local buckling under small loads in some sections and that these elements were then capable of carrying higher loads even after local buckling of the members.

In the United States, the first edition of the Specification for the Design of Light Gage Steel Structural Members was published by the American Iron and Steel Institute (AISI) in 1946 (AISI, 1946). The first Allowable Stress Design (ASD) Specification was based on the research work sponsored by AISI at Cornell University under the direction of late Professor George Winter since 1939. As a result of this work, George Winter is now considered the grandfather of cold-formed steel design. The ASD Specification was subsequently revised in 1956, 1960, 1962, 1968, 1980, and 1986 to reflect the technical developments and the results of continued research at Cornell and other universities (Yu et al., 1996). In 1991, AISI published the first edition of the Load and Resistance Factor Design Specification developed at University of Missouri of Rolla and Washington University under the directions of Wei-Wen Yu and Theodore V. Galambos (AISI, 1991). Both ASD and LRFD Specifications were combined into a single specification in 1996 (AISI, 1996).

In 2001, the first edition of the North American Specification for the Design of Cold-Formed Steel Structural Members was developed by a joint effort of the AISI Committee on Specifications, the Canadian Standards Association (CSA) Technical Committee on Cold-Formed Steel Structural Members, and Camara Nacional de la Industria del Hierro y del Acero (CANACERO) in Mexico (AISI, 2001). It included the ASD and LRFD methods for the United States and Mexico together with the Limit States Design (LSD) method for Canada. This North American Specification has been accredited by the American National Standard Institute (ANSI) as an ANSI Standard to supersede the 1996 AISI Specification and the 1994 CSA Standard. Following the successful use of the 2001 edition of the North American Specification for six years, it was revised and expanded in 2007.

This updated specification includes new and revised design provisions with the additions of the Direct Strength Method in Appendix 1 and the Second-Order Analysis of structural systems in Appendix 2.

In addition to the AISI specifications, the American Iron and Steel Institute has also published commentaries on various editions of the specifications, design manuals, framing design standards, various design guides, and design aids for using cold-formed steel.

### International codes and standards

The United States, Mexico and Canada use the North American Specification for the Design of Cold-Formed Steel Structural Members, document number AISI S100-2007. Member states of the European Union use section 1-3 of the Eurocode 3 (EN 1993) for the design of cold formed steel members. Other nations utilize various design specifications, many based on AISI S-100, as adopted by the building codes listed below. Another list of international cold-formed steel codes and standards is maintained (and can be edited with permission) at Cold-Formed Steel Codes Around the World.

### Africa

Ethiopia Building Codes: EBCS-1 Basis of design and actions on structures EBCS-3 Design of steel structures

South Africa Specification: SANS 10162 - The Structural Use of Steel: Part 2 - Limit-state design of cold-formed steelwork Building code: National Building Regulations of South Africa

## Americas

United States Specification: North American Specification for the Design of Cold-Formed Steel Structural Members, document number AISI S100-2007 published by the American Iron and Steel Institute in October 2007. Building Code: IBC and/or NFPA may be enforced, but both reference AISI S100.

Canada Specification: North American Specification for the Design of Cold-Formed Steel Structural Members, document number CAN/CSA S136-07 as published by Canadian Standards Association which is the same as AISI S100 except for the cover. Building Code: The National Building Code of Canada is the model code adopted with amendments by individual Provinces and Territories. The Federal government is outside the jurisdiction of the Provincial/Territorial authority but usually defers to the legislated requirements within the Province/Territory of the building site.

Brazil Specification: NBR 14762:2001 Dimensionamento de estruturas de aço constituídas por perfis formados a frio - Procedimento (Cold-formed steel design - Procedure, last update 2001) and NBR 6355:2003 Perfis estruturais de aço formados a frio - Padronização (Cold-formed steel structural profiles, last update 2003) Building Code: ABNT - Associação Brasileira de Normas Técnicas (www.abnt.org.br)

Chile NCH 427 - suspended because it was written in the 1970s. Cold-formed steel sections were based in part on AISI (U.S). The local Institute for Building code INN has specified in recent Codes for seismic design that designers must use the last edition of the AISI Specification for cold formed steel and the AISC for hot rolled, in their original versions in English until some traduced adaption will be issued here .

Argentina CIRSOC 303 for Light Steel Structures where cold formed steel is included. That Specification, now more than 20 years old, is being replaced by a new one, which will be, in general, an adaption of the current AISI one. The former CIRSOC 303 was an adaption of the Canadian code of that time. At this time CIRSOC 303 was very old, now CIRSOC 301 is in revolution to be aligned with the American codes (LRFD design). In the near future both codes will be aligned also in designations and therminology.

## Asia

Philippines National Structural Code of the Philippines (NSCP) 2010, Volume 1 Buildings, Towers, and other Vertical Structures, Chapter 5 Part 3 Design of Cold-Formed Steel Structural Members is based on AISI S100-2007

India Specification:IS:801, Indian standard code of practice for use of cold-formed light gauge steel structural members in general building construction, Bureau of Indian Standards, New Delhi (1975).

China Specification: Technical Code of Cold-formed Thin-wall Steel Structures Building Code: GB 50018-2002 (current version)

Japan Specification: Design Manual of Light-gauge Steel Structures Building Code: Technical standard notification No.1641 concerning light-gauge steel structures

Malaysia Malaysia uses British Standard BS5950, especially BS5950:Part 5; AS4600 (from Australia) is also referenced.

## Europe

EU Countries Specification: EN 1993-1-3 (same as Eurocode 3 part 1-3), Design of steel structures - Cold formed thin gauge members and sheeting. Each European country will get its own National Annex Documents (NAD).

Germany Specification: German Committee for Steel Structures (DASt), DASt-Guidelines 016: 1992: Calculation and design of structures with thin-walled cold-formed members; In German Building Code: EN 1993-1-3: 2006 (Eurocode 3 Part 1-3): Design of steel structures – General rules – Supplementary rules for cold-formed members and sheeting; German version prEN 1090 2: 2005 (prEN 1090 Part 2; Draft): Execution of steel structures and aluminium structures – Technical requirements for the execution of steel structures; German version EN 10162: 2003: Cold-rolled steel sections – Technical delivery conditions – Dimensional and cross-sectional tolerances; German version

Italy Specification: UNI CNR 10022 (National Document) EN 1993-1-3 (Not compulsory)

United Kingdom Eurocode for cold-formed steel in the UK. BS EN 1993-1-3:2006: Eurocode 3. Design of steel structures. General rules.

## Oceania

Australia Specification: AS/NZS 4600 AS/NZS 4600:2005 Similar to NAS 2007 but includes high strength steels such as G550 for all sections. (Greg Hancock) Building Code: Building Code of Australia (National document) calls AS/NZS 4600:2005

New Zealand Specification: AS/NZS 4600 (same as Australia)

Common section profiles and applications

In building construction there are basically two types of structural steel: hot-rolled steel shapes and cold-formed steel shapes. The hot rolled steel shapes are formed at elevated temperatures while the cold-formed steel shapes are formed at room temperature. Cold-formed steel structural members are shapes commonly manufactured from steel plate, sheet metal or strip material. The manufacturing process involves forming the material by either press-braking or cold roll forming to achieve the desired shape.

When steel is formed by press-braking or cold rolled forming, there is a change in the mechanical properties of the material by virtue of the cold working of the metal. When a steel section is cold-formed from flat sheet or strip the yield strength, and to a lesser extent the ultimate strength, are increased as a result of this cold working, particularly in the bends of the section.

Some of the main properties of cold formed steel are as follows:

- Lightness in weight

- High strength and stiffness

- Ease of prefabrication and mass production

- Fast and easy erection and installation

- Substantial elimination of delays due to weather

- More accurate detailing

- Non shrinking and non creeping at ambient temperatures

- No formwork needed

- Termite-proof and rot proof

- Uniform quality

- Economy in transportation and handling

- Non combustibility

- Recyclable material

- Panels and decks can provide enclosed cells for conduits.

A broad classification of the cold-formed shapes used in the construction industry can be made as individual structural framing members or panels and decks.

Some of the popular applications and the preferred sections are:

- Roof and wall systems (industrial, commercial, and agricultural buildings)

- Steel racks for supporting storage pallets

- Structural members for plane and space trusses

- Frameless Stressed skin structures: Corrugated sheets or sheeting profiles with stiffened edges are used for small structures up to a 30 ft clear span with no interior framework

CFS Decking

CFS purlins

CFS X-braced wall system

CFS stud/girt wall connection

## Typical stress–strain properties

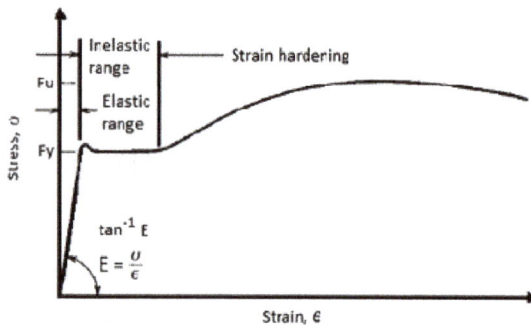
Figure 1

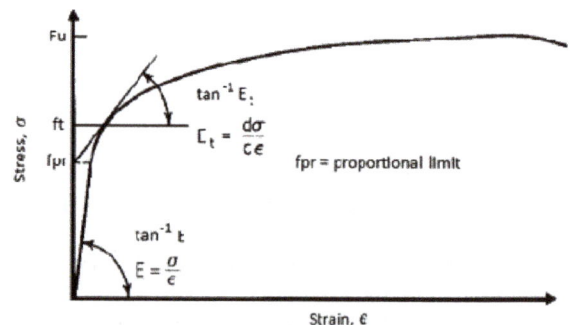
Figure 2

A main property of steel, which is used to describe its behavior, is the stress–strain graph. The stress–strain graphs of cold-formed steel sheet mainly fall into two categories. They are sharp yielding and gradual yielding type illustrated below in Fig.1 and Fig.2, respectively.

These two stress–strain curves are typical for cold-formed steel sheet during tension test. The second graph is the representation of the steel sheet that has undergone the cold-reducing (hard rolling) during manufacturing process, therefore it does not exhibit a yield point with a yield plateau. The initial slope of the curve may be lowered as a result of the prework. Unlike Fig.1, the stress–strain relationship in Fig.2 represents the behavior of annealed steel sheet. For this type of steel, the yield point is defined by the level at which the stress–strain curve becomes horizontal.

Cold forming has the effect of increasing the yield strength of steel, the increase being the consequence of cold working well into the strain-hardening range. This increase is in the zones where the material is deformed by bending or working. The yield stress can be assumed to have been increased by 15% or more for design purposes. The yield stress value of cold-formed steel is usually between 33ksi and 80ksi. The measured values of Modulus of Elasticity based on the standard methods usually range from 29,000 to 30,000 ksi (200 to 207 GPa). A value of 29,500 ksi

(203 GPa) is recommended by AISI in its specification for design purposes. The ultimate tensile strength of steel sheets in the sections has little direct relationship to the design of those members. The load-carrying capacities of cold-formed steel flexural and compression members are usually limited by yield point or buckling stresses that are less than the yield point of steel, particularly for those compression elements having relatively large flat-width ratios and for compression members having relatively large slenderness ratios. The exceptions are bolted and welded connections, the strength of which depends not only on the yield point but also on the ultimate tensile strength of the material. Studies indicate that the effects of cold work on formed steel members depend largely upon the spread between the tensile and the yield strength of the virgin material.

## Ductility Criteria

Ductility is defined as "an extent to which a material can sustain plastic deformation without rupture." It is not only required in the forming process but is also needed for plastic redistribution of stress in members and connections, where stress concentration would occur. The ductility criteria and performance of low-ductility steels for cold-formed members and connections have been studied by Dhalla, Winter, and Errera at Cornell University. It was found that the ductility measurement in a standard tension test includes local ductility and uniform ductility. Local ductility is designated as the localized elongation at the eventual fracture zone. Uniform ductility is the ability of a tension coupon to undergo sizeable plastic deformations along its entire length prior to necking. This study also revealed that for the different ductility steels investigated, the elongation in 2-in. (50.8-mm) gage length did not correlate satisfactorily with either the local or the uniform ductility of the material. In order to be able to redistribute the stresses in the plastic range to avoid premature brittle fracture and to achieve full net-section strength in a tension member with stress concentrations, it is suggested that:

- The minimum local elongation in a - 1–2 in. (12.7-mm) gauge length of a standard tension coupon including the neck be at least 20%.

- The minimum uniform elongation in a 3-in. (76.2-mm) gauge length minus the elongation in a 1-in. (25.4-mm) gage length containing neck and fracture be at least 3%.

- The tensile-strength-to-yield-point ratio $F_u/F_y$ be at least 1.05.

## Weldability

Weldability refers to the capacity of steel to be welded into a satisfactory, crack free, sound joint under fabrication conditions without difficulty. Welding is possible in cold-formed steel elements, but it shall follow the standards given in AISI S100-2007, Section E.

1.When thickness less than or equal to 3/16" (4.76mm):

The various possible welds in cold formed steel sections, where the thickness of the thinnest element in the connection is 3/16" or less are as follows

- o   Groove Welds in Butt joints

- o   Arc Spot Welds

- o   Arc Seam Welds

- o   Fillet Welds

- o   Flare Groove Welds

2. When thickness greater than or equal to 3/16" (4.76mm):

Welded connections in which thickness of the thinnest connected arc is greater than 3/16" (4.76mm) shall be in accordance with **ANSI/AISC-360**. The weld positions are covered as per **AISI S100-2007** (Table E2a)

Minimum material thickness recommended for welding connections

| Application | Shop or Field fabrication | Electrode method | Suggested minimum CFS thickness |
|---|---|---|---|
| CFS to Structural steel | Field-fabrication | Stick-welding | 54 mils to 68 mils |
| CFS to Structural steel | Shop-fabrication | Stick-welding | 54 mils to 68 mils |
| CFS to CFS | Field-fabrication | Stick-welding | 54 mils to 68 mils |
| CFS to CFS | Field-fabrication | Wire-fed MIG (Metal Inert Gas) welding | 43 mils to 54 mils |
| CFS to CFS | Shop-fabrication | Wire-fed MIG (Metal Inert Gas) welding | 33 mils |

## Application in Buildings

## Cold-formed Steel Framing

Cold-formed steel framing (CFSF) refers specifically to members in light-frame building construction that are made entirely of sheet steel, formed to various shapes at ambient temperatures. The most common shape for CFSF members is a lipped channel, although "Z", "C", tubular, "hat" and other shapes and variations have been used. The building elements that are most often framed with cold-formed steel are floors, roofs, and walls, although other building elements and both structural and decorative assemblies may be steel framed.

Although cold-formed steel is used for several products in building construction, framing products are different in that they are typically used for wall studs, floor joists, rafters, and truss members. Examples of cold-formed steel that would not be considered framing includes metal roofing, roof and floor deck, composite deck, metal siding, and purlins and girts on metal buildings.

Framing members are typically spaced at 16 or 24 inches on center, with spacing variations lower and higher depending upon the loads and coverings. Wall members are typically vertical lipped channel "stud" members, which fit into unlipped channel "track" sections at the top and bottom. Similar configurations are used for both floor joist and rafter assemblies, but in a horizontal application for floors, and a horizontal or sloped application for roof framing. Additional elements of the framing system include fasteners and connectors, braces and bracing, clips and connectors.

In North America, member types have been divided into five major categories, and product nomenclature is based on those categories.

- S members are lipped channels, most often used for wall studs, floor joists, and ceiling or roof rafters.

- T members are unlipped channels, which are used for top and bottom plates (tracks) in walls, and rim joists in floor systems. Tracks also form the heads and sills of windows, and typically cap the top and bottom of boxed- or back-to-back headers.

- U members are unlipped channels that have a smaller depth than tracks, but are used to brace members, as well as for ceiling support systems.

- F members are "furring" or "hat" channels, typically used horizontally on walls or ceilings.

- L members are angles, which in some cases can be used for headers across openings, to distribute loads to the adjacent jamb studs.

In high-rise commercial and multi-family residential construction, CFSF is typically used for interior partitions and support of exterior walls and cladding. In many mid-rise and low-rise applications, the entire structural system can be framed with CFSF.

## Connectors and Fasteners in Framing

Connectors are used in cold-formed steel construction to attach members (i.e. studs, joists) to each other or to the primary structure for the purpose of load transfer and support. Since an assembly is only as strong as its weakest component, it is important to engineer each connection so that it meets specified performance requirements. There are two main connection types, **Fixed and Movement-Allowing** (Slip). Fixed connections of framing members do not allow movement of the connected parts. They can be found in axial-load bearing walls, curtain walls, trusses, roofs, and floors. Movement-Allowing connections are designed to allow deflection of the primary structure in the vertical direction due to live load, or in the horizontal direction due to wind or seismic loads, or both vertical and horizontal directions. One application for a vertical movement-allowing connection is to isolate non-axial load bearing walls (drywall) from the vertical live load of the structure and to prevent damage to finishes. A common clip for this application is an L-shaped top-of-wall clip for walls that are infill between floors. These clips have slots perpendicular to the bend in the clip. Another common clip is the bypass clip for walls that bypass outside the edge of the floor structure. When these clips are L-shaped, the have slots that are parallel to the bend in the clip. If the structure is in an active seismic zone, vertical and horizontal movement-allowing connections may be used to accommodate both the vertical deflection and horizontal drift of the structure.

Connectors may be fastened to cold-formed steel members and primary structure using welds, bolts, or self-drilling screws. These fastening methods are recognized in the American Iron and Steel Institute (AISI) 2007 North American Specification for the Design of Cold-Formed Steel Structural Members, Chapter E. Other fastening methods, such as clinching, power actuated fasteners (PAF), mechanical anchors, adhesive anchors and structural glue, are used based on manufacturer's performance-based tests.

**Hot-rolled versus cold-rolled steel and the influence of annealing**

|  |  | Hot rolled | Cold rolled |
|---|---|---|---|
| Material properties | Yielding strength | The material is not deformed; there is no initial strain in the material, hence yielding starts at actual yield value as the original material. | The yield value is increased by 15%–30% due to prework (initial deformation). |
|  | Modulus of elasticity | 29,000 ksi | 29,500 ksi |
|  | Unit weight | Unit weight is comparatively huge. | It is much smaller. |
|  | Ductility | More ductile in nature. | Less ductile. |
| Design |  | Most of the time, we consider only the global buckling of the member. | Local buckling, Distortional Buckling, Global Buckling have to be considered. |
| Main uses |  | Load bearing structures, usually heavy load bearing structures and where ductility is more important ( Example Seismic prone areas) | Application in many variety of loading cases. This includes building frames, automobile, aircraft, home appliances, etc. Use limited in cases where high ductility requirements. |
| Flexibility of shapes |  | Standard shapes are followed. High value of unit weight limits the flexibility of manufacturing wide variety of shapes. | Any desired shape can be molded out of the sheets. The light weight enhances its variety of usage. |
| Economy |  | High Unit weight increases the overall cost – material, lifting, transporting, etc. It is difficult to work with (e.g. connection). | Low unit weight reduces the cost comparatively. Ease of construction (e.g. connection). |
| Research possibilities |  | In the advanced stages at present. | More possibilities as the concept is relatively new and material finds wide variety of applications. |

Annealing, also described in the earlier section, is part of the manufacturing process of cold-formed steel sheet. It is a heat treatment technique that alters the microstructure of the cold-reducing steel to recover its ductility.

## Alternative Design Methods

The Direct Strength Method (DSM) is an alternative method of design located in Appendix 1 of the *North American Specification for the Design of Cold-formed Steel Structural Members* 2007 (AISI S100-07). DSM may be used in lieu of the Main Specification for determining nominal member capacities. Specific advantages include the absence of effective width and iterations, while only using known gross-sectional properties. An increase in prediction confidence stems from forced compatibility between section flanges and web throughout elastic buckling analysis. This increase in prediction accuracy for any section geometry provides a solid basis for rational analysis extension and encourages cross-sectional optimization. Either DSM or the main specification can be used with confidence as the $\Phi$ or $\Omega$ factors have been designed to insure that both methods are accurate. Currently, DSM only provides solutions for beams and columns and has to be used in conjunction with the main specification for a complete design.

Rational analysis is permitted when using optimized cold form shapes that are outside of the scope of the main specification and are not pre-qualified for DSM use. These non-pre-qualified sections use the factors of safety of $\Phi$ and $\Omega$ associated with rational analysis. The result of the rational analysis times the appropriate factor of safety will be used as the design strength of the section.

Several situations may exist where a rational analysis application of DSM can be used. In general these would include: (1) determining the elastic buckling values and (2) using the DSM equations in Appendix 1 to determine nominal flexural and axial capacities, Mn and Pn. The premise of DSM itself is an example of rational analysis. It uses elastic buckling results to determine the ultimate strength through the use of empirical strength curves. This provides designers with a method for performing a rational analysis in a number of unique situations.

In some cases the rational analysis extension to DSM may be as simple as dealing with an observed buckling mode that is difficult to identify and making a judgment call as to how to categorize the mode. But it could also be used to allow an engineer to include the effects of moment gradients, the influence of different end conditions, or the influence of torsion warping on all buckling modes.

There are currently no provisions within the DSM that pertain to shear, web crippling, holes in members, or strength increases due to the cold work of forming. Research on several of these topics has been completed or is in the process of being completed and should be included in the next update of the AISI Specification. DSM is also limited in determining strength for sections in which very slender elements are used. This is due to the strength of a cross section being predicted as a whole with DSM instead of using the effective width method of the specification which breaks the cross section up into several effective elements. One slender element will cause low strength with DSM, which is not the case with the current specification method. The finite strip method using CUFSM is the most commonly used approach to determine the elastic buckling loads. The program also limits DSM because holes cannot be considered, loads have to be uniform along the member, only simply supported boundary conditions are considered, and the buckling modes interact and cannot be easily distinguishable in some cases.

## Sheet Metal

Sheet metal is metal formed by an industrial process into thin, flat pieces. It is one of the fundamental forms used in metalworking and it can be cut and bent into a variety of shapes. Countless everyday objects are constructed with sheet metal. Thicknesses can vary significantly; extremely thin thicknesses are considered foil or leaf, and pieces thicker than 6 mm (0.25 in) are considered plate.

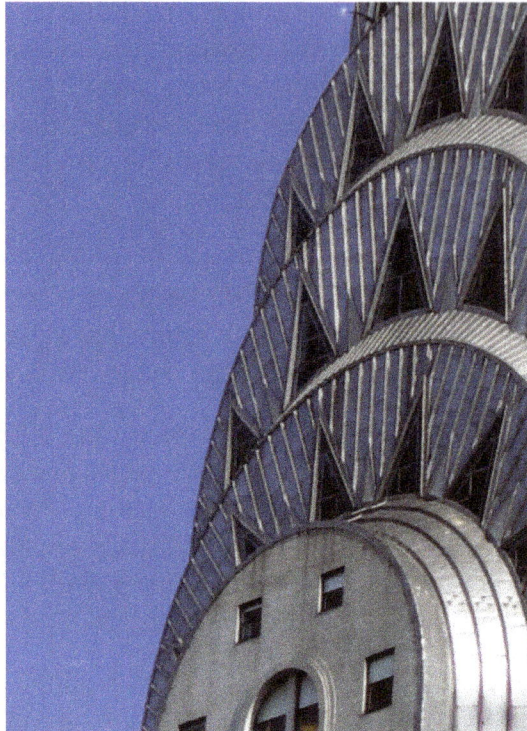

Sheets of *Nirosta* stainless steel cover the Chrysler Building

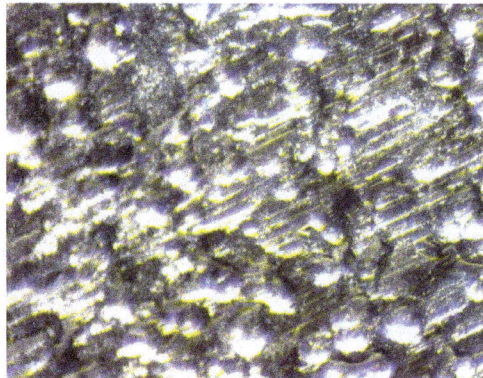

Microscopic close-up of mild steel sheet metal.

Sheet metal is available in flat pieces or coiled strips. The coils are formed by running a continuous sheet of metal through a roll slitter.

The thickness of sheet metal is in the USA commonly specified by a traditional, non-linear measure known as its gauge. The larger the gauge number, the thinner the metal. Commonly used steel sheet metal ranges from 30 gauge to about 7 gauge. Gauge differs between ferrous (iron based) metals and nonferrous metals such as aluminum or copper; copper thickness, for example is measured in ounces (and represents the thickness of 1 ounce of copper rolled out to an area of 1 square foot). In the rest of the world the sheet metal thickness is given in millimeters.

There are many different metals that can be made into sheet metal, such as aluminum, brass, copper, steel, tin, nickel and titanium. For decorative uses, important sheet metals include silver, gold, and platinum (platinum sheet metal is also utilized as a catalyst.)

Sheet metal is used for car bodies, airplane wings, medical tables, roofs for buildings (architecture) and many other applications. Sheet metal of iron and other materials with high magnetic permeability, also known as laminated steel cores, has applications in transformers and electric machines. Historically, an important use of sheet metal was in plate armor worn by cavalry, and sheet metal continues to have many decorative uses, including in horse tack. Sheet metal workers are also known as "tin bashers" (or "tin knockers"), a name derived from the hammering of panel seams when installing tin roofs.

## Materials

### Stainless Steel

Grade 304 is the most common of the three grades. It offers good corrosion resistance while maintaining formability and weldability. Available finishes are #2B, #3, and #4. Grade 303 is not available in sheet form.

Grade 316 possesses more corrosion resistance and strength at elevated temperatures than 304. It is commonly used for pumps, valves, chemical equipment, and marine applications. Available finishes are #2B, #3, and #4.

Grade 410 is a heat treatable stainless steel, but it has a lower corrosion resistance than the other grades. It is commonly used in cutlery. The only available finish is dull.

Grade 430 is popular grade, low cost alternative to serie 300's grades. Used when high corrosion resistance is not a primary criteria. Common grade for appliance products, often with a brushed finish.

### Aluminium

Aluminum is also a popular metal used in sheet metal due to its flexibility, wide range of options, cost effectiveness, and other properties. The four most common aluminium grades available as sheet metal are 1100-H14, 3003-H14, 5052-H32, and 6061-T6.

Grade 1100-H14 is commercially pure aluminium, highly chemical and weather resistant. It is ductile enough for deep drawing and weldable, but has low strength. It is commonly used in chemical processing equipment, light reflectors, and jewelry.

Grade 3003-H14 is stronger than 1100, while maintaining the same formability and low cost. It is corrosion resistant and weldable. It is often used in stampings, spun and drawn parts, mail boxes, cabinets, tanks, and fan blades.

Grade 5052-H32 is much stronger than 3003 while still maintaining good formability. It maintains high corrosion resistance and weldability. Common applications include electronic chassis, tanks, and pressure vessels.

Grade 6061-T6 is a common heat-treated structural aluminium alloy. It is weldable, corrosion resistant, and stronger than 5052, but not as formable. It loses some of its strength when welded. It is used in modern aircraft structures.

## Gauge

Use of Gauge to designate sheet metal thickness is discouraged by numerous international standards organizations. For Example, ASTM states in specification ASTM A480-10a "The use of gage number is discouraged as being an archaic term of limited usefulness not having general agreement on meaning."

Manufacturers' Standard Gauge for Sheet Steel is based on an average weight of 41.82 lb (18.96 kg) per square foot per inch thick. Gauge is defined differently for ferrous (iron-based) and non-ferrous metals (e.g., aluminium and brass).

| Standard sheet metal gauges | | | | | | |
|---|---|---|---|---|---|---|
| Gauge | U.S. standard for sheet and plate iron and steel decimal inch (mm) | Steel inch (mm) | Galvanized steel inch (mm) | Stainless steel inch (mm) | Aluminium inch (mm) | Zinc inch (mm) |
| 0000000 | 0.5000 (12.70) | ...... | ...... | ...... | ...... | ...... |
| 000000 | 0.4688 (11.91) | ...... | ...... | ...... | ...... | ...... |
| 00000 | 0.4375 (11.11) | ...... | ...... | ...... | ...... | ...... |
| 0000 | 0.4063 (10.32) | ...... | ...... | ...... | ...... | ...... |
| 000 | 0.3750 (9.53) | ...... | ...... | ...... | ...... | ...... |
| 00 | 0.3438 (8.73) | ...... | ...... | ...... | ...... | ...... |
| 0 | 0.3125 (7.94) | ...... | ...... | ...... | ...... | ...... |
| 1 | 0.2813 (7.15) | ...... | ...... | ...... | ...... | ...... |
| 2 | 0.2656 (6.75) | ...... | ...... | ...... | ...... | ...... |
| 3 | 0.2500 (6.35) | 0.2391 (6.07) | ...... | ...... | ...... | 0.006 (0.15) |
| 4 | 0.2344 (5.95) | 0.2242 (5.69) | ...... | ...... | ...... | 0.008 (0.20) |
| 5 | 0.2188 (5.56) | 0.2092 (5.31) | ...... | ...... | ...... | 0.010 (0.25) |
| 6 | 0.2031 (5.16) | 0.1943 (4.94) | ...... | ...... | 0.162 (4.1) | 0.012 (0.30) |
| 7 | 0.1875 (4.76) | 0.1793 (4.55) | ...... | 0.1875 (4.76) | 0.1443 (3.67) | 0.014 (0.36) |
| 8 | 0.1719 (4.37) | 0.1644 (4.18) | 0.1681 (4.27) | 0.1719 (4.37) | 0.1285 (3.26) | 0.016 (0.41) |
| 9 | 0.1563 (3.97) | 0.1495 (3.80) | 0.1532 (3.89) | 0.1563 (3.97) | 0.1144 (2.91) | 0.018 (0.46) |
| 10 | 0.1406 (3.57) | 0.1345 (3.42) | 0.1382 (3.51) | 0.1406 (3.57) | 0.1019 (2.59) | 0.020 (0.51) |
| 11 | 0.1250 (3.18) | 0.1196 (3.04) | 0.1233 (3.13) | 0.1250 (3.18) | 0.0907 (2.30) | 0.024 (0.61) |

| 12 | 0.1094 (2.78) | 0.1046 (2.66) | 0.1084 (2.75) | 0.1094 (2.78) | 0.0808 (2.05) | 0.028 (0.71) |
| 13 | 0.0938 (2.38) | 0.0897 (2.28) | 0.0934 (2.37) | 0.094 (2.4) | 0.072 (1.8) | 0.032 (0.81) |
| 14 | 0.0781 (1.98) | 0.0747 (1.90) | 0.0785 (1.99) | 0.0781 (1.98) | 0.0641 (1.63) | 0.036 (0.91) |
| 15 | 0.0703 (1.79) | 0.0673 (1.71) | 0.0710 (1.80) | 0.07 (1.8) | 0.057 (1.4) | 0.040 (1.0) |
| 16 | 0.0625 (1.59) | 0.0598 (1.52) | 0.0635 (1.61) | 0.0625 (1.59) | 0.0508 (1.29) | 0.045 (1.1) |
| 17 | 0.0563 (1.43) | 0.0538 (1.37) | 0.0575 (1.46) | 0.056 (1.4) | 0.045 (1.1) | 0.050 (1.3) |
| 18 | 0.0500 (1.27) | 0.0478 (1.21) | 0.0516 (1.31) | 0.0500 (1.27) | 0.0403 (1.02) | 0.055 (1.4) |
| 19 | 0.0438 (1.11) | 0.0418 (1.06) | 0.0456 (1.16) | 0.044 (1.1) | 0.036 (0.91) | 0.060 (1.5) |
| 20 | 0.0375 (0.95) | 0.0359 (0.91) | 0.0396 (1.01) | 0.0375 (0.95) | 0.0320 (0.81) | 0.070 (1.8) |
| 21 | 0.0344 (0.87) | 0.0329 (0.84) | 0.0366 (0.93) | 0.034 (0.86) | 0.028 (0.71) | 0.080 (2.0) |
| 22 | 0.0313 (0.80) | 0.0299 (0.76) | 0.0336 (0.85) | 0.031 (0.79) | 0.025 (0.64) | 0.090 (2.3) |
| 23 | 0.0281 (0.71) | 0.0269 (0.68) | 0.0306 (0.78) | 0.028 (0.71) | 0.023 (0.58) | 0.100 (2.5) |
| 24 | 0.0250 (0.64) | 0.0239 (0.61) | 0.0276 (0.70) | 0.025 (0.64) | 0.02 (0.51) | 0.125 (3.2) |
| 25 | 0.0219 (0.56) | 0.0209 (0.53) | 0.0247 (0.63) | 0.022 (0.56) | 0.018 (0.46) | ...... |
| 26 | 0.0188 (0.48) | 0.0179 (0.45) | 0.0217 (0.55) | 0.019 (0.48) | 0.017 (0.43) | ...... |
| 27 | 0.0172 (0.44) | 0.0164 (0.42) | 0.0202 (0.51) | 0.017 (0.43) | 0.014 (0.36) | ...... |
| 28 | 0.0156 (0.40) | 0.0149 (0.38) | 0.0187 (0.47) | 0.016 (0.41) | 0.0126 (0.32) | ...... |
| 29 | 0.0141 (0.36) | 0.0135 (0.34) | 0.0172 (0.44) | 0.014 (0.36) | 0.0113 (0.29) | ...... |
| 30 | 0.0125 (0.32) | 0.0120 (0.30) | 0.0157 (0.40) | 0.013 (0.33) | 0.0100 (0.25) | ...... |
| 31 | 0.0109 (0.28) | 0.0105 (0.27) | 0.0142 (0.36) | 0.011 (0.28) | 0.0089 (0.23) | ...... |
| 32 | 0.0102 (0.26) | 0.0097 (0.25) | ...... | ...... | ...... | ...... |
| 33 | 0.0094 (0.24) | 0.0090 (0.23) | ...... | ...... | ...... | ...... |

| 34 | 0.0086 (0.22) | 0.0082 (0.21) | ...... | ...... | ...... | ...... |
| 35 | 0.0078 (0.20) | 0.0075 (0.19) | ...... | ...... | ...... | ...... |
| 36 | 0.0070 (0.18) | 0.0067 (0.17) | ...... | ...... | ...... | ...... |
| 37 | 0.0066 (0.17) | 0.0064 (0.16) | ...... | ...... | ...... | ...... |
| 38 | 0.0063 (0.16) | 0.0060 (0.15) | ...... | ...... | ...... | ...... |

## Tolerances

During the rolling process the rollers bow slightly, which results in the sheets being thinner on the edges. The tolerances in the table and attachments reflect current manufacturing practices and commercial standards and are not representative of the Manufacturer's Standard Gauge, which has no inherent tolerances.

| Steel sheet metal tolerances | | | |
|---|---|---|---|
| Gauge | Nominal [in (mm)] | Max [in (mm)] | Min [in (mm)] |
| 10 | 0.1345 (3.42) | 0.1405 (3.57) | 0.1285 (3.26) |
| 11 | 0.1196 (3.04) | 0.1256 (3.19) | 0.1136 (2.89) |
| 12 | 0.1046 (2.66) | 0.1106 (2.81) | 0.0986 (2.50) |
| 14 | 0.0747 (1.90) | 0.0797 (2.02) | 0.0697 (1.77) |
| 16 | 0.0598 (1.52) | 0.0648 (1.65) | 0.0548 (1.39) |
| 18 | 0.0478 (1.21) | 0.0518 (1.32) | 0.0438 (1.11) |
| 20 | 0.0359 (0.91) | 0.0389 (0.99) | 0.0329 (0.84) |
| 22 | 0.0299 (0.76) | 0.0329 (0.84) | 0.0269 (0.68) |
| 24 | 0.0239 (0.61) | 0.0269 (0.68) | 0.0209 (0.53) |
| 26 | 0.0179 (0.45) | 0.0199 (0.51) | 0.0159 (0.40) |
| 28 | 0.0149 (0.38) | 0.0169 (0.43) | 0.0129 (0.33) |

| Aluminium sheet metal tolerances | | |
|---|---|---|
| **Thickness [in (mm)]** | **Sheet width** | |
| | **36 (914.4) [in (mm)]** | **48 (1,219) [in (mm)]** |
| 0.018–0.028 (0.46–0.71) | 0.002 (0.051) | 0.0025 (0.064) |
| 0.029–0.036 (0.74–0.91) | 0.002 (0.051) | 0.0025 (0.064) |
| 0.037–0.045 (0.94–1.14) | 0.0025 (0.064) | 0.003 (0.076) |
| 0.046–0.068 (1.2–1.7) | 0.003 (0.076) | 0.004 (0.10) |
| 0.069–0.076 (1.8–1.9) | 0.003 (0.076) | 0.004 (0.10) |
| 0.077–0.096 (2.0–2.4) | 0.0035 (0.089) | 0.004 (0.10) |
| 0.097–0.108 (2.5–2.7) | 0.004 (0.10) | 0.005 (0.13) |
| 0.109–0.125 (2.8–3.2) | 0.0045 (0.11) | 0.005 (0.13) |
| 0.126–0.140 (3.2–3.6) | 0.0045 (0.11) | 0.005 (0.13) |
| 0.141–0.172 (3.6–4.4) | 0.006 (0.15) | 0.008 (0.20) |
| 0.173–0.203 (4.4–5.2) | 0.007 (0.18) | 0.010 (0.25) |
| 0.204–0.249 (5.2–6.3) | 0.009 (0.23) | 0.011 (0.28) |

| Stainless steel sheet metal tolerances | | |
|---|---|---|
| **Thickness [in (mm)]** | **Sheet width** | |
| | **36 (914.4) [in (mm)]** | **48 (1,219) [in (mm)]** |
| 0.017–0.030 (0.43–0.76) | 0.0015 (0.038) | 0.002 (0.051) |
| 0.031–0.041 (0.79–1.04) | 0.002 (0.051) | 0.003 (0.076) |
| 0.042–0.059 (1.1–1.5) | 0.003 (0.076) | 0.004 (0.10) |
| 0.060–0.073 (1.5–1.9) | 0.003 (0.076) | 0.0045 (0.11) |
| 0.074–0.084 (1.9–2.1) | 0.004 (0.10) | 0.0055 (0.14) |
| 0.085–0.099 (2.2–2.5) | 0.004 (0.10) | 0.006 (0.15) |
| 0.100–0.115 (2.5–2.9) | 0.005 (0.13) | 0.007 (0.18) |
| 0.116–0.131 (2.9–3.3) | 0.005 (0.13) | 0.0075 (0.19) |
| 0.132–0.146 (3.4–3.7) | 0.006 (0.15) | 0.009 (0.23) |
| 0.147–0.187 (3.7–4.7) | 0.007 (0.18) | 0.0105 (0.27) |

## Forming processes

## Bending

The equation for estimating the maximum bending force is,

$$F_{Max} = k\frac{TLt^2}{W},$$

where $k$ is a factor taking into account several parameters including friction. $T$ is the ultimate tensile strength of the metal. $L$ and $t$ are the length and thickness of the sheet metal, respectively. The variable $W$ is the open width of a V-die or wiping die.

## Curling

## Decambering

## Deep Drawing

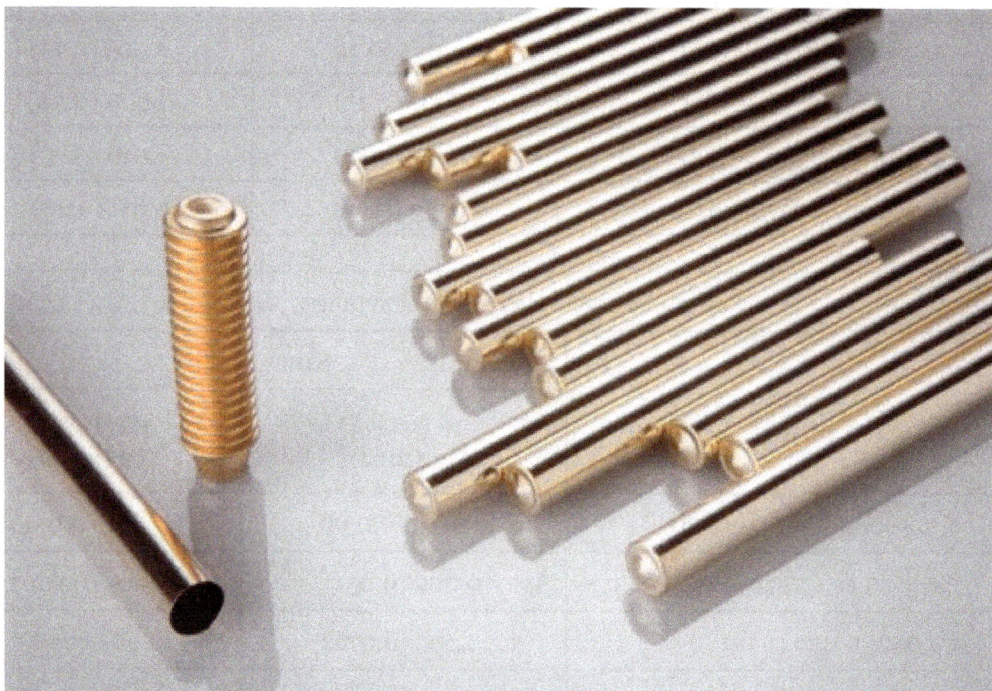

Example of deep drawn part

Drawing is a forming process in which the metal is stretched over a form or die. In deep drawing the depth of the part being made is more than half its diameter. Deep drawing is used for making automotive fuel tanks, kitchen sinks, two-piece aluminum cans, etc. Deep drawing is generally done in multiple steps called draw reductions. The greater the depth the more reductions are required. Deep drawing may also be accomplished with fewer reductions by heating the workpiece, for example in sink manufacture.

In many cases, material is rolled at the mill in both directions to aid in deep drawing. This leads to a more uniform grain structure which limits tearing and is referred to as "draw quality" material.

## Expanding

Expanding is a process of cutting or stamping slits in alternating pattern much like the stretcher bond in brickwork and then stretching the sheet open in accordion-like fashion. It is used in applications where air and water flow are desired as well as when light weight is desired at cost of a solid flat surface. A similar process is used in other materials such as paper to create a low cost packing paper with better supportive properties than flat paper alone.

## Hydroforming

Hydroforming is a process that is analogous to deep drawing, in that the part is formed by stretching the blank over a stationary die. The force required to do so is generated by the direct application of extremely high hydrostatic pressure to the workpiece or to a bladder that is in contact with the workpiece, rather than by the movable part of a die in a mechanical or hydraulic press. Unlike deep drawing, hydroforming usually does not involve draw reductions—the piece is formed in a single step.

## Incremental Sheet Forming

## Ironing

## Laser Cutting

Sheet metal can be cut in various ways, from hand tools called tin snips up to very large powered shears. With the advances in technology, sheet metal cutting has turned to computers for precise cutting. Many sheet metal cutting operations are based on computer numerically controlled (CNC) laser cutting or multi-tool CNC punch press.

CNC laser involves moving a lens assembly carrying a beam of laser light over the surface of the metal. Oxygen, nitrogen or air is fed through the same nozzle from which the laser beam exits. The metal is heated and burnt by the laser beam, cutting the metal sheet. The quality of the edge can be mirror smooth and a precision of around 0.1 mm (0.0039 in) can be obtained. Cutting speeds on thin 1.2 mm (0.047 in) sheet can be as high as 25 m (82 ft) a minute. Most of the laser cutting systems use a $CO_2$ based laser source with a wavelength of around 10 μm; some more recent systems use a YAG based laser with a wavelength of around 1 μm.

## Photochemical Machining

Photochemical machining, also known as photo etching, is a tightly controlled corrosion process which is used to produce complex metal parts from sheet metal with very fine detail. The photo etching process involves photo sensitive polymer being applied to a raw metal sheet. Using CAD designed photo-tools as stencils, the metal is exposed to UV light to leave a design pattern, which is developed and etched from the metal sheet.

## Perforating

Perforating is a cutting process that punches multiple small holes close together in a flat workpiece. Perforated sheet metal is used to make a wide variety of surface cutting tools, such as the surform.

## Press Brake Forming

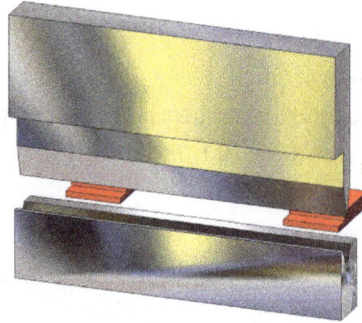

Forming metal on a pressbrake

This is a form of bending used to produce long, thin sheet metal parts. The machine that bends the metal is called a press brake. The lower part of the press contains a V-shaped groove called the die. The upper part of the press contains a punch that presses the sheet metal down into the v-shaped die, causing it to bend. There are several techniques used, but the most common modern method is "air bending". Here, the die has a sharper angle than the required bend (typically 85 degrees for a 90 degree bend) and the upper tool is precisely controlled in its stroke to push the metal down the required amount to bend it through 90 degrees. Typically, a general purpose machine has an available bending force of around 25 tonnes per metre of length. The opening width of the lower die is typically 8 to 10 times the thickness of the metal to be bent (for example, 5 mm material could be bent in a 40 mm die). The inner radius of the bend formed in the metal is determined not by the radius of the upper tool, but by the lower die width. Typically, the inner radius is equal to 1/6 of the V-width used in the forming process.

The press usually has some sort of back gauge to position depth of the bend along the workpiece. The backgauge can be computer controlled to allow the operator to make a series of bends in a component to a high degree of accuracy. Simple machines control only the backstop, more advanced machines control the position and angle of the stop, its height and the position of the two reference pegs used to locate the material. The machine can also record the exact position and pressure required for each bending operation to allow the operator to achieve a perfect 90 degree bend across a variety of operations on the part.

## Punching

Punching is performed by placing the sheet of metal stock between a punch and a die mounted in a press. The punch and die are made of hardened steel and are the same shape. The punch is sized to be a very close fit in the die. The press pushes the punch against and into the die with enough force to cut a hole in the stock. In some cases the punch and die "nest" together to create a depression in the stock. In progressive stamping a coil of stock is fed into a long die/punch set with many stages. Multiple simple shaped holes may be produced in one stage, but complex holes are created in multiple stages. In the final stage, the part is punched free from the "web".

A typical CNC turret punch has a choice of up to 60 tools in a "turret" that can be rotated to bring any tool to the punching position. A simple shape (e.g., a square, circle, or hexagon) is cut directly from the sheet. A complex shape can be cut out by making many square or rounded cuts around the perimeter. A punch is less flexible than a laser for cutting compound shapes, but faster for repetitive shapes (for example, the grille of an air-conditioning unit). A CNC punch can achieve 600 strokes per minute.

A typical component (such as the side of a computer case) can be cut to high precision from a blank sheet in under 15 seconds by either a press or a laser CNC machine..

## Roll Forming

A continuous bending operation for producing open profiles or welded tubes with long lengths or in large quantities.

## Rolling

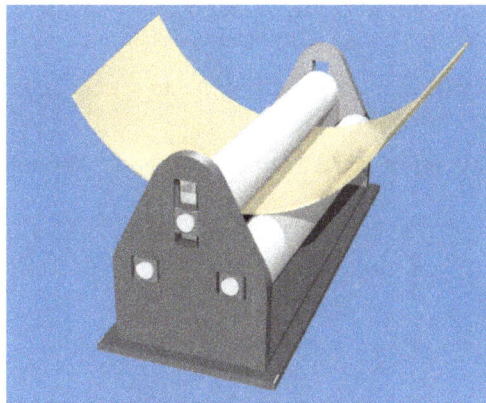

Bending sheet metal with rollers

## Spinning

Spinning is used to make tubular (axis-symmetric) parts by fixing a piece of sheet stock to a rotating form (mandrel). Rollers or rigid tools press the stock against the form, stretching it until the stock takes the shape of the form. Spinning is used to make rocket motor casings, missile nose cones, satellite dishes and metal kitchen funnels.

## Stamping

Stamping includes a variety of operations such as punching, blanking, embossing, bending, flanging, and coining; simple or complex shapes can be formed at high production rates; tooling and equipment costs can be high, but labor costs are low.

Alternatively, the related techniques repoussé and chasing have low tooling and equipment costs, but high labor costs..

## Water Jet Cutting

A water jet cutter, also known as a waterjet, is a tool capable of a controlled erosion into metal or other materials using a jet of water at high velocity and pressure, or a mixture of water and an abrasive substance.

# References

- Hesse, Rayner, W. (2007). Jewelrymaking through History: an Encyclopedia. Greenwood Publishing Group. p. 56. ISBN 0-313-33507-9.

- Possehl, Gregory L. (1996). Mehrgarh in Oxford Companion to Archaeology, Brian Fagan (Ed.). Oxford University Press. ISBN 0-19-507618-4

- Degarmo, E. Paul; Black, J T.; Kohser, Ronald A. (2003), Materials and Processes in Manufacturing (9th ed.), Wiley, ISBN 0-471-65653-4.

- Landes, David. S. (1969). The Unbound Prometheus: Technological Change and Industrial Development in Western Europe from 1750 to the Present. Cambridge, New York: Press Syndicate of the University of Cambridge. p. 91. ISBN 0-521-09418-6.

- Todd, Robert H.; Allen, Dell K.; Alting, Leo (1994), Manufacturing Processes Reference Guide, Industrial Press Inc., pp. 300–304, ISBN 0-8311-3049-0.

- Green, Archie (1993). Wobblies, pile butts, and other heroes : laborlore explorations. Urbana u.a.: Univ. of Illinois Press. p. 20. ISBN 9780252019630. Retrieved 14 July 2015.

- Parker, Dana T. Building Victory: Aircraft Manufacturing in the Los Angeles Area in World War II, p. 20, 85, Cypress, CA, 2013. ISBN 978-0-9897906-0-4.

- Rowlett, Ross (26 July 2002). "Sheet Metal Thickness Gauges". University of North Carolina at Chapel Hill. Retrieved 21 June 2013.

# Permissions

All chapters in this book are published with permission under the Creative Commons Attribution Share Alike License or equivalent. Every chapter published in this book has been scrutinized by our experts. Their significance has been extensively debated. The topics covered herein carry significant information for a comprehensive understanding. They may even be implemented as practical applications or may be referred to as a beginning point for further studies.

We would like to thank the editorial team for lending their expertise to make the book truly unique. They have played a crucial role in the development of this book. Without their invaluable contributions this book wouldn't have been possible. They have made vital efforts to compile up to date information on the varied aspects of this subject to make this book a valuable addition to the collection of many professionals and students.

This book was conceptualized with the vision of imparting up-to-date and integrated information in this field. To ensure the same, a matchless editorial board was set up. Every individual on the board went through rigorous rounds of assessment to prove their worth. After which they invested a large part of their time researching and compiling the most relevant data for our readers.

The editorial board has been involved in producing this book since its inception. They have spent rigorous hours researching and exploring the diverse topics which have resulted in the successful publishing of this book. They have passed on their knowledge of decades through this book. To expedite this challenging task, the publisher supported the team at every step. A small team of assistant editors was also appointed to further simplify the editing procedure and attain best results for the readers.

Apart from the editorial board, the designing team has also invested a significant amount of their time in understanding the subject and creating the most relevant covers. They scrutinized every image to scout for the most suitable representation of the subject and create an appropriate cover for the book.

The publishing team has been an ardent support to the editorial, designing and production team. Their endless efforts to recruit the best for this project, has resulted in the accomplishment of this book. They are a veteran in the field of academics and their pool of knowledge is as vast as their experience in printing. Their expertise and guidance has proved useful at every step. Their uncompromising quality standards have made this book an exceptional effort. Their encouragement from time to time has been an inspiration for everyone.

The publisher and the editorial board hope that this book will prove to be a valuable piece of knowledge for students, practitioners and scholars across the globe.

# Index

www.ingramcontent.com/pod-product-compliance
Lightning Source LLC
Chambersburg PA
CBHW061304190326
41458CB00011B/3759